Recording Studio Technology, Maintenance, and Repairs

Recording Studio Technology, Maintenance, and Repairs

Tom McCartney

McGraw-Hill
New York   Chicago   San Francisco   Lisbon
London   Madrid   Mexico City   Milan   New Delhi
San Juan   Seoul   Singapore   Sydney   Toronto

Contents

Preface

My goal in life is to provide a service to the industry that I love by way of teaching and providing as many of the tips and tricks that I can. I am a 25-year veteran of the industry as a repairperson, engineering technician, manager, salesperson and consultant for some of the major brand name companies of the industry. The knowledge that I have gained needs to be passed along and not kept secret.

The music and efforts of the people involved are always inspiring. Seeing the customer's satisfaction as they begin to operate their equipment after repairs or upgrades is a truly rewarding experience and makes it all worthwhile.

This book is designed to help remove some of the mysteries and unnecessary repair bills that add up to quite sizeable amounts over time. Communication and understanding is the secondary purpose. The person who reads and uses this book as a guide will understand better how to communicate with their service person using some of the technical terminology associated with this business.

The contents in this book are to provide a better understanding of the inner workings of equipment and some of the maintenance and minor repairs for these items where applicable. This book shows the practices which I use and is not a substitute for the manufacturers recommended service procedures in their service manual.

Always remember what a good tradeperson does (measure twice, cut once). In other words, where possible, read the service information (if available) on the item before attempting any repairs or disassembly.

Tom McCartney

Acknowledgments

I would like to express thanks to my customers who allowed me to use their equipment in the photographs and, on occasion, some of my students, who assisted me in preparing the photos. I would also like to express my most sincere thanks to Sam, my friend and customer, who convinced me to start teaching and writing about the things that I have learned through my years in the business. In his words I quote " It would be such a waste, not to pass this stuff on to someone!" Well Sam, look what you started!

This is the first of what I hope will be small series of books on the procedures, tips, and tricks of getting equipment repaired or at least back up and running without losing the session. The musicians / customers are paying our bills and we need to keep things as smooth and positive as possible for them. Hopefully, these books will be of great value to those who need to do it themselves occasionally. Studio Survival Skills!

I am also inspired by my late father, who would help young people starting out in life.

Also many thanks to those who took the time to teach me throughout the years.

Many thanks to McGraw-Hill and Associated Publishing companies for their patience and help in getting this launched.

CALCULATIONS AND EXPLANATIONS

CHAPTER ONE

Electricity—What Is It?

One way to describe electricity is that it is a force that transfers energy from one point to another in the form of atomic pressure. Electricity behaves like water in a hose. A battery acts in the same way as a pump that creates pressure. This pressure produces a flow if the valve is turned on. The flow can be used if it is attached to something like a sprinkler, which uses the flow to operate a set of gears that rotate the sprayer back and forth.

Let us take this description a step further and compare electricity to a cement mixer truck. The truck has an oil pump that creates pressure. The driver turns on a valve and the pressure from the pump creates flow through the line to the oil-driven motor, which turns the large drum of cement on the truck.

This process is similar to a battery or generator producing a voltage, which, when the switch is turned on, creates current flow and may operate a fan. The pressure from the pump is like the voltage from the battery. The gallons per minute of flow from the pump are like the amperes of current from the battery. The load on the pump, being the resistance of the drum to rotate when it is full of cement, is similar to the resistance in a circuit. The higher the resistance, the lower the flow or current. If we turn up the pressure or voltage, then we force more flow or current through the circuit, just like the driver revving up the engine in the cement mixer to turn the drum.

Figure 1-1 illustrates an example of this flow in a liquid system.

Figure 1-2 illustrates an electrical circuit that behaves in a similar way as the liquid system.

FIGURE 1-1
A liquid system

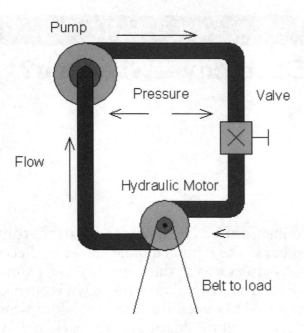

Pump

Pressure

Valve

Flow

Hydraulic Motor

Belt to load

FIGURE 1-2
An example of an
electrical circuit

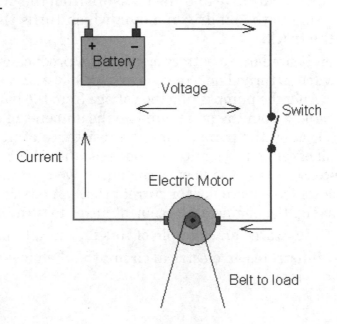

Battery

Voltage

Switch

Current

Electric Motor

Belt to load

CHAPTER TWO

Basic Electronics: DC Voltage, Current, and Resistance

The best way to think of voltage, current, and resistance is by considering the way water pressure behaves. Please refer to the basic formulas from time to time.

Voltage can be thought of as a form of pressure. If you have a water pump that produces 100 pounds of pressure, this is similar to voltage. A battery produces voltage as a result of a chemical reaction that takes place inside it. The acid is trying to break down a high atomic weight metal and sacrifice it to a low atomic weight material like carbon. This chemical reaction builds up a voltage and stops when the pressure (voltage) reaches the atomic difference between the two different types of materials. A battery can be made of something as simple as a lemon, a nail, and a piece of copper. Sticking the nail and piece of copper into the lemon will create a pressure or voltage between them because the acid allows the two metals to react with each other. So, think of voltage as pressure, which is waiting to be released by sending current or flow somewhere through a conductor similar to a pipe. A battery delivers DC current, which is a direct current that flows in a single direction from the battery or generator's negative terminal through the bulb or load and back to the battery or generator's positive terminal.

Current can be thought of as flow. If the pressure is in the water hose and you pull the nozzle trigger on the end, water flies out of the hose and goes somewhere. Current in a wire is similar, with the exception that it needs to return to the other side of the source. If you have a battery and want to send current through a piece of wire, you must connect it to both sides of the battery. The danger is that if you just connect a wire from the positive (+) terminal to the negative (−) one you will have very little or no voltage difference between the two terminals. This is because you will short the two terminals out with the wire, and all the available current will rush to the other end in a rather dangerous way. The current will go from the negative through the wire to the positive and accomplish nothing except perhaps heating up the wire and battery. This would waste power, just like a pressure hose not connected to anything and just spraying water instead of moving a sprinkler unit, for example.

The sprinkler unit could be considered a load on the hose just like a motor or light bulb on a battery. If you have water pressure in a hose and you attach it to the sprinkler, the sprinkler will operate and begin to move around, doing some work like moving back and forth while absorbing some of the energy from the pressure that was available. If you have a battery and you connect a light bulb between the positive and negative terminals, the bulb absorbs some of the energy and lights up, providing enough pressure existed to heat the filament. The filament has a positive temperature coefficient and the resistance increases with temperature, so when the bulb is first switched on, an amount equal to 100 times the operating current is used until the filament heats up. You would need to use a bulb that would have enough resistance to create the required amount of heat to make the filament glow white-hot. Then you would have a match between the voltage and current to be able to work properly together. This resistance is like having a restriction in the flow of the hose. It limits the amount of current.

If you had a water hose that had 100 pounds of pressure but only a tiny pump, you probably would not be able to operate the lawn sprinkler because there would not be enough pressure to sustain a flow strong enough to operate the sprinkler. If you had a tiny sprinkler and hooked it to the tiny pump, it may operate. The tiny sprinkler would represent a higher resistance.

The same goes for the battery. You certainly cannot start your car with a flashlight battery, but you could operate a 12-volt flashlight for a long time on a car battery (see Figure 2-1). The pump or battery must be able to keep the pressure up while delivering the required amount of flow. Thus, the battery would have to be big enough to be able to provide the necessary current for the load. Picture trying to operate a cordless drill on a 9-volt camera battery. It may spin slowly, but not for long because the chemicals and materials would deliver a small amount of current that would be consumed in a very short time.

A formula exists for this: Voltage in volts divided by resistance in ohms is equal to current in amperes. To summarize, pressure is voltage, flow is current, and restriction is resistance. Please see the formula pages.

The amount of work done by the load is calculated or measured in watts. Watts can be calculated by multiplying the voltage by the current. Thus, a 12-volt car battery delivering 3 amps of current to a headlamp will be producing 36 watts of energy. Because the headlamp is consuming the energy, the bulb must be rated at 36 watts.

Early Edison DC hydroelectric generators stored extra power in batteries kept in rooms for peak use, and they usually had a steam engine as a backup generator. These systems were local to villages and wealthy houses, and they could not be connected far from the generator and batteries.

These systems could not transmit power for long distances due to losses in the wires. AC power was created to be able to transport large amounts of power over long distances.

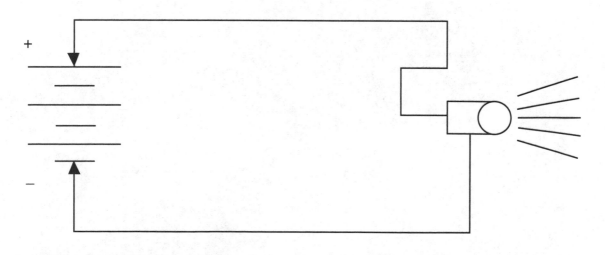

FIGURE 2-1
Battery and Light Bulb

CHAPTER THREE

Basic Electronics: AC Voltage, Current, and Reactance

AC voltage and AC current are varying values of DC voltage and current that reverse in direction continuously. As with DC, think of the voltage as pressure and the current as flow. The pressure is positive for one half-cycle and then negative for the other half. The first half-cycle may try to push flow out of the hose (wire) and the next half-cycle tries to suck flow back in, completing the cycle when it goes back to zero on its way toward the positive terminal.

The rate at which this change occurs is called frequency and is measured in Hertz (Hz). A 60 Hz 120-volt power line voltage starts at zero, rises to a maximum peak of 170 volts, falls through zero, and ends up at -170 volts. This gives a difference of 340 volts. You're probably thinking, what about 120 volts? During the rise and fall of the voltages, the actual voltage is sometimes more, and it is mostly less than 120 volts. The average or *root means squared* (RMS) value is 120 from the 340-volt peak-to-peak value; 340 divided by 2 equals 170 divided by 1.44 is equal to 120.

That is why getting a shock from AC is different from DC at the same voltage. The DC shock is like getting bitten by a small animal, whereas an AC shock is like getting bitten and shaken by a large dog (speaking from experience, of course). The changing voltage tends to make muscles contract and then release at the same rate as the power, causing an out-of-control grasp on the wire until the body can throw itself off. In a DC shock, you are mostly thrown off instantly by a single, large contraction of muscles. In either case, it is highly painful and dangerous.

The AC waveforms are sinusoidal, which resemble parabolic curves, as shown in Figure 3-1.

The calculations for watts, amperes, ohms, and volts remain the same same in AC circuits as in DC circuits for most practical and quick applications. The ratings for some equipment are in *volt-amperes* (VA), which is nearly the same as watts. It is a term that allows for efficiency and a power factor, which we do not need to discuss at this time.

AC power was created also because of the need to change the generating station's values from time to time. DC voltage cannot be changed cheaply because transformers only react to changing current. DC does not change; it is either there or not. Imagine a power station trying to send you power at 120-volts DC from the nearest generating station. Your house probably consumes 6,000 watts of power in the summer at meal or laundry time with the air conditioner on. A town of 200 homes would need 1.2 million watts; 1,200,000 divided by 120 equals 10,000 am-

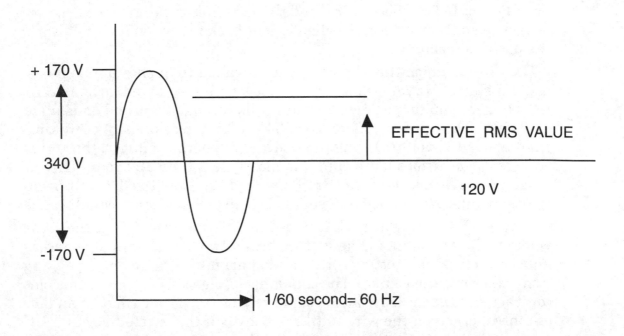

FIGURE 3-1
Sinusoidal AC waveforms

peres. The alternative would be to send high voltage and low current, and branch off with motor-generator units to convert the power down to usable voltages. This would be ridiculously expensive. Motors and generators are expensive and require maintenance and repairs.

Instead of low voltage and high current, let's et's try 60 Hz cycles of AC power at a high voltage and a low current. Generate 240,000 volts to send to the area. It will drop to 30,000 volts at a station, go to around 6,000 volts at a substation, and finally down to 120 and 220 at the transformer in the basement of the office or on the street corner for the houses. The reason the voltage is so high from the generating station is because of the wattage required by the customers. If you are a generating station and need to sell 100,000 watts of power to a town 100 miles away, you would need wires the size of small tunnels to carry that much current if you sent your power at 120 volts; 100,000 divided by 120 volts is equal to 833 amperes.

If you send the power needed at 240,000 volts, the wires can be about 1 inch in diameter because you are sending under 30 amps; 1,000,000 watts divided by 250,000 volts equals 4 amperes for that town. That is why power tools require heavy extension cords to be able to carry the 10 or 12 amperes of current to make up the required power for the circular saw. One unit of horsepower needs 750 watts, so a skill saw will need 10 amperes to operate. Therefore, you need to carry 10 amperes in that cable. The longer the extension cord is, the larger it needs to be in order to minimize heat and losses over the distance.

The transformer is a form of reactor because it converts electrical energy in the primary winding or coil into magnetic energy and then back to electrical energy in the secondary winding. It transfers power from one circuit to another without the circuits being physically connected together. Thanks to these devices, far fewer people get a shock from appliances and radios than in earlier times.

The frequency that North America runs on is 60 cycles, whereas Europe runs at 50 cycles. North America started out at 30 and went to 60 to get rid of flickering lights and humming devices. We use this accurate 60-cycle power, which is synchronized all over North America, to keep time with our electric clocks and other devices. The modern transformers and AC motors are designed for the best efficiency at 60 Hz so that some devices overheat a little when taken to Europe and operated at 50 Hz. These devices would be items such as power amplifiers and tape recorders, where the power supplies are mostly analog transformer style. Guitar amps would also hum and have heating trouble due to a lack of efficiency in the power supplies at lower-frequency power without larger transformers. Because of the lower the line frequency, the more magnetic material is needed in the core of the transformer.

CHAPTER FOUR

Basic Formulas for Calculations

Ohm's Law

This is one of the most common formulas in the electronics world. It provides the basic answers for approximating current consumption in various circuits. The basic values are as follows:

I represents the current in amperes.

E represents the voltage in volts.

R represents the resistance in ohms.

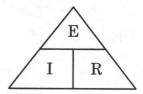

To find the current, divide the voltage by the resistance. To find the resistance, divide the voltage by the current. To find the voltage, multiply the current by the resistance.

To find the impedance or resistance of multiple loads, let's say you have 2 sets of speakers, each at 8 ohms, and you want to know what the overall impedance will be on the amplifier per channel. Let's call the first speaker set R1 and the second R2. The formula would read as follows:

$$(R1 \times R2)/(R1 + R2).$$

$$8 \times 8 = 64 ... ... 8 + 8 = 16 ... .. 64 / 16 = 4$$

The overall impedance of the 2 pairs of speakers would be 4 ohms.

Let's try 2 unmatched pairs, one at 8 ohms and one at 4 ohms:

$$8 \times 4 = 32 \ldots\ 8 + 4 = 12 \ldots 32 / 12 = 2.66$$

Therefore, the impedance of the 2 sets of speakers together would be 2.66 ohms per channel.

Let's try 3 pairs of speakers, one pair at 8 ohms, one pair at 6 ohms, and one pair at 4 ohms:

$$8 \times 6 = 48,\ 8 + 6 = 14,\ 48/14 = 3.43,\ 3.43 \times 4 = 13.72,\ 3.43 + 4 = 7.43$$
$$13.72/7.43 = 1.85$$

The overall impedance per channel would be 1.85 ohms.

Another way to do this is

$$1/RT = 1/(1/R1 + 1/R2 + 1/R3)$$
$$1/(1/8\ +\ 1/6\ + 1/4\ ) = 1/.125 + .166 + .25 = 1/.541 = 1.848$$

TABLE 4-1   Ohm's Law Calculations Table

Watts (W)	Amperes (A)
$E \times I$	E/R
E^2/R	W/E
$I^2 \times R$	$\sqrt{W/R}$

Volts (V)	Ohms (R)
$I \times R$	E/I
W/I	E^2/W
$\sqrt{W \times R}$	W/I^2

Power: DC and Basic AC

P represents power in watts. The other symbols are listed earlier.

$$P = (E) \times (I), \ P = (I^2) \times R, \ P = (E^2)/R$$

DC or basic AC power calculations can be used for most simple applications. As an example, let's say you have a multitrack recorder and a mixing console on the same circuit, and you are wondering if they will trip the circuit breaker. The multitrack is rated for 800 watts and the mixing console is rated for 400 watts. The current required will be calculated as follows:

$$400 + 800 = 1200 \text{ watts}, \ 1200 = 120 \text{ volts} \times (I), \ (I) = 1200/120 = 10 \text{ amperes}$$

Therefore, a 15-ampere circuit will be adequate for these devices. As a rule of thumb, a 50 percent cushion is a good idea to have on hand, so I would not add anything else to that circuit. Just for curiosity, let's calculate the resistance of the circuit: 120 volts divided by 10 amperes equals 12 ohms of AC resistance or impedance.

The voltage you measure is in RMS and is rated this way as an equivalent to DC power. The actual voltage is 1.414 times the RMS, so 120 volts RMS is actually 170 volts peak. The peak to peak would be 170 volts × 2 = 340 volts.

This is also why some engineers like to use *peak program meters* (PPM) instead of *Volume Units* (VU) meters in their consoles. PPM gives a peak reading instead of an average so that the transients can be seen (see metering in Chapter 8). A PPM meter reads peak voltages, whereas the VU reads RMS.

Going back to the speakers, let's figure out how much power will be zsent to each speaker from above when the amp is delivering nearly full power at 100 watts:

$$100 = E^2/2.66, \ E^2 = 100 \times 2.66, \ E = \sqrt{(100 \times 2.66)}, \ E = 16.31 \text{ volt}$$

The current from the amplifier is 16.31 volts divided by 2.66 ohms, which equals 6.13 amps. Let's test this:

$$16.31 \text{ volts} \times 6.13 \text{ amps} = 100 \text{ watts}$$

How about that, it works! The power through the 8 ohm speaker is 16.32^2. Divided by 8, it is 33.3 watts. The power through the 4 ohm speaker is 16.32^2. Divided by 4, it is 66.6 watts.

The 4 ohm speaker will therefore be 20 percent louder than the 8 ohm speaker. This may come as a surprise because twice as much power is going through the 4 ohm speaker, so you may think it should be double. Sorry, but it doesn't work that way. To double the volume, you need 10 times the power. If you double the power, you only get one-fifth louder in volume. That phenomenon is called a decilinear response known as the decibel, which is equal to $10 \log \times (P_i/P_o)$, where P_i is power in and P_o is power out. Most amplifiers are considered to have 1 milliwatt input or .001 W. (See the following section.)

CHAPTER FIVE

Decibels

The decibel is a unit of volume or power measurement. It can relate to gain, volume power, signal strength, and so on.

The variations are as follows:

dBu (most often used in audio) .775 volts (unspecified impedance)

dBj 1/1000 volts

dBk 1000 watts

dBm Milliwatts at 600 ohms impedance

dBs Japanese equivalent to dBm

dBv 1 volt (no longer used)

dBw 1 watt

dBvg Voltage gain

dBrap Decibels above a reference acoustical power of 10^{-16} in watts

VU A reading in dB relative to the nominal operating level. Where the reference level is +4dBu, the VU meter would read "0" at about 60% of full deflection.

This is also called 4 dB of lead. The −80 dB range would typically be the background noise of an effects device or a midpriced console with channels on but faders down. The +40 range would be the level of a small power amplifier.

Most high-quality mixers can put out a maximum of 24 dB at the balanced outputs while amplifying a signal as weak as −80 dB from a microphone. That is a gain structure of over 100 dB, which has amplified the signal by over 100,000 times its original level.

A Decibel to Voltage level Table showing the relationship between the two.

V = Volts, MV = millivolts 1/1,000 V, uV = microvolts 1/1,000,000 V

TABLE 5-1   A Typical Reference

dB	Voltage	dB	Voltage	dB	Voltage	dB	Voltage
−80	75 µV	−70	245 µV	−60	775 µV	−50	2.45 mV
−40	7.75 mV	−30	24.5 mV	−20	77.5 mV	−10	.245 V
0	.775 V	+4	1.24 V	+6	1.55 V	+8	1.95 V
+10	2.45 V	+20	7.75 V	+30	24.5 V	+40	77.5 V

To calculate the gain in dB of an amplification stage, you would multiply 20 by (log (V_{out}/V_{in})). Calculating the attenuation of a stage would require the same formula except the value of dB would be negative.

To calculate the power gain in dB of a power amplifier, you would multiply 10 by (log (P_{out}/P_{in})). If you applied 0 dB to an amplifier and wanted to know its gain, you would measure its output AC volts into a known value load. Let's use 4 ohms. Our calculations would be as follows: Po is E^2/R, and P in is 1 milliwatt or .001 watts.

At 0dB input, we measured 27 volts:

$$27 \times 27 = 729, \ 729/4 = 182.25 \text{ watts}$$

$$(P_o/P_i) \log \times 10 \text{ (where } P_o \text{ is Power out and } P_i \text{ is Power in)}$$

$$(182.25/.001)\log 10 \ = 182,250 \ \log 10 = 5.26066 \times 10 = 52.6 \text{ dB gain}$$

CHAPTER SIX

Advanced Formulas for Calculations

Reactance

To find out the impedance a coil or capacitor has on a circuit, two basic formulas can be used. Reactance is like resistance, but it changes with frequency. For coils, it increases with frequency. For capacitors, reactance decreases with frequency.

The technically correct symbol for an AC load is Z, as opposed to R for DC resistance. Z represents what is called impedance. XC is capacitive reactance, and XL is inductive reactance. Z is impedance and is measured in ohms.

Capacitive reactance: $XC = 1 / (2 \times \pi \times$ the frequency in Hz $\times$ the capacitor's value in farads)

Inductive reactance: $XL = 2 \times \pi \times$ the frequency in Hz $\times$ the coil's value in henrys

So, if the frequency is 10 Hz and the capacitor is 50 microfarad (μF), what is the reactance of the capacitor? The answer is 310 ohms.

By transposing the formula, we could find the coil value that would provide the same reactance or impedance at the same frequency. The value would be 5 henrys, which would be a rather large coil.

When these two reactances match, you have a resonant circuit. This is how radios tune to different stations by selecting the desired frequency.

$$FR = \frac{1}{(2 \times 3.14 \times \sqrt{(L \times C)})}$$

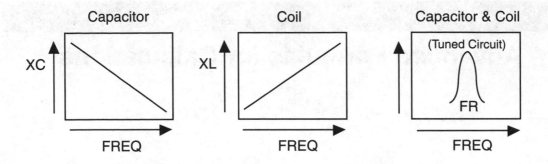

FIGURE 6-1
Reactions of Capacitors and Coils, Separate and connected together.

FR is the resonant frequency, $\sqrt{}$ is square root, π is 3.14, L is the coil value in henrys, and C is the capacitor value in farads (see Figure 6-1).

Attenuators

Unbalanced L Type

$$R1 = Z \times ((K - 1)/K)\ R2 = Z \times (1/(K - 1))$$

R1 is the series resistor, R2 is the shunt resistor, and Z is the impedance that is required in ohms.

Unbalanced T Type

$$R1 \text{ and } R2 = ((K - 1)/(K + 1)) \times Z$$
$$R3 = (K/(K^2 - 1)) \times 2 \times Z \quad \text{where } Z \text{ is the impedance}$$

Balanced U Type

$$R1 = (Z1/2\ S) \times ((KS - 1)/K)$$

$$R2 = (Z1/S) \times (1/(K - S)) \quad \text{where } S = \sqrt{Z1/Z2}$$

R1 is the input series resistors, R2 is the shunting resistor, and Z is the impedance required in ohms.

Lattice-Type Attenuator

$$R1 = ((K - 1) / (K + 1)) \times Z \quad R2 = ((K + 1) / (K - 1)) \times Z$$

where Z is the impedance required in ohms

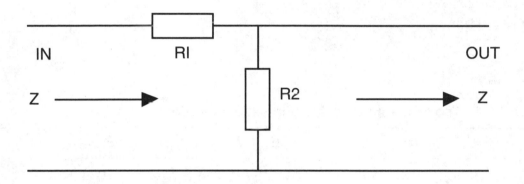

FIGURE 6-2
Unbalanced L type

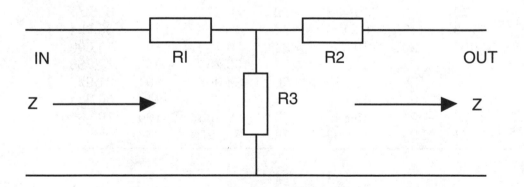

FIGURE 6-3
Unbalanced T type

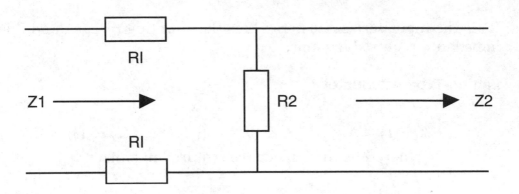

FIGURE 6-4
Balanced U type

TABLE 6-1   The K factors for calculating attenuator loss

dB	K	dB	K	dB	K	dB	K
0.05	1.0058	12	3.9811	26	19.953	40	100.00
0.1	1.0116	13	4.4668	27	22.387	45	177.83
0.5	1.0593	14	5.0119	28	25.119	50	316.23
1.0	1.1220	15	5.6234	29	28.184	55	562.34
2.0	1.2589	16	6.3096	30	31.623	60	1000.0
3.0	1.4125	17	7.0795	31	35.481	65	1778.3
4.0	1.5849	18	7.9433	32	39.811	70	3162.3
5.0	1.7783	19	8.9125	33	44.668	75	5623.4
6.0	1.9953	20	10.000	34	50.119	80	10,000
7.0	2.2387	21	11.2202	35	56.234	85	17,783
8.0	2.5119	22	12.589	36	63.096	90	31,623
9.0	2.8184	23	14.125	37	70.795	95	56,234
10	3.1623	24	15.849	38	79.433	100	100,000
11	3.5481	25	17.783	39	89.125	——	

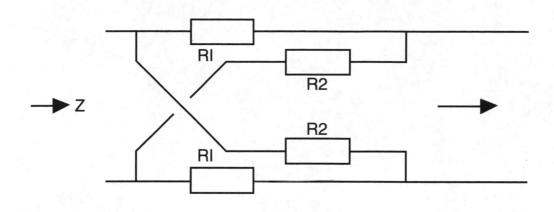

FIGURE 6-5
Lattice-type attenuator

CHAPTER SEVEN

Wire Sizes and Maximum Current for Each Wire

AWG is the wire size, and Mils is the circular area, maximum amps in open air, maximum amps in a confined space (chassis) (see Table 7-1).

Another way of looking at wire size is shown in Table 7-2.

The larger the wire, the lower the resistance, and thus the more current it can carry over longer distances without creating heat. Heat is energy lost in the wire that did not get to the desired destination.

TABLE 7-1

| Wire Size | | Copper conductor | Max. | |
AWG	Circular Mils	(100° C) nominal R Ohms per 1000 ft	Amps Air	Conf.
32	63.2	188.0	0.53	0.32
30	100.5	116.0	0.86	0.52
28	159.8	72.0	1.4	0.83
26	254.1	45.2	2.2	1.3
24	404.0	28.4	3.5	2.1
22	642.4	22.0	7.0	5.0
20	1022	13.7	11	7.5
18	1624	6.50	16	10
16	2583	5.15	22	13
14	4107	3.20	32	17
12	6530	2.02	41	23
10	10,380	1.31	55	33
8	16,510	0.734	73	46
6	26,250	0.459	101	60
4	41,740	0.290	135	80
2	66,370	0.185	181	100
1	83,690	0.151	211	125
0	105,500	0.117	245	150
00	133,100	0.092	283	175
000	167,800	0.074	328	200

TABLE 7-2

Wire gage	Diameter (in)	Feet per ohm
0	0.368	3448
02	0.292	1362
04	0.232	3448
08	0.147	1362
10	0.1019	763
12	0.0403	495
14	0.0581	312
16	0.0508	194
18	0.0403	156
20	0.032	95
22	0.025	59
24	0.02	37
26	0.016	23
28	0.018	18
30	0.013	14
32	0.008	8.5
36	0.005	2.25
38	0.0040	1.5

CHAPTER EIGHT

Technical Power and Current Requirements

Clean AC power can be one of the most important factors in a quiet studio. The last thing your client wants to hear is an annoying buzzing sound from the monitors while they are listening to their artistic creation or, worse, having noise recorded on their masters during overdubs.

If you are not sure about having the proper technical power or you have noise in your system, you can test for this in a few quick ways. . You will need an AC voltmeter, an AC extension cord, and some electrical tape. With the voltmeter set to read *AC 600-volt range,* insert a probe into the small vertical slot in an extension cord, and tape the remaining exposed metal shaft of the same probe. Plug the extension cord into the outlet near the console. Being careful not to touch any metal part with your hand, insert the other probe into the small vertical slot in other outlet and check for voltage. If 230 volts are used, the circuit being measured is on a different phase. If little or no voltage exists, the circuit is on the same phase. If you measure around 115 volts, a problem has occurred, such as a dead circuit or an error in the wiring, and an electrician should be called soon.

If you are careful, you can check the light dimmers. Shut off the breaker or remove the fuse while opening the dimmer box. Hang the dimmer unit out of the box without letting it touch anything metal. Turn the breaker back on or reinstall the fuse, and measure the terminals or wires of the dimmer while the dimmer is turned up full. Then turn off the power and reinstall the dimmer.

What you are looking for is to see if the other outlets are on the same phase as the console. The outlets that show no voltage can be used for other audio devices, whereas those that show 230 volts should be used for computers, coffeepots, appliances, and other nonaudio devices. Dimmers and lights should measure 230 volts, which would indicate that they are also on a different phase from the audio. The lighting system is the most important to have on a different phase and it should read 230 volts relative to the console and all other audio systems (see Figures 8-1 and 8-2).

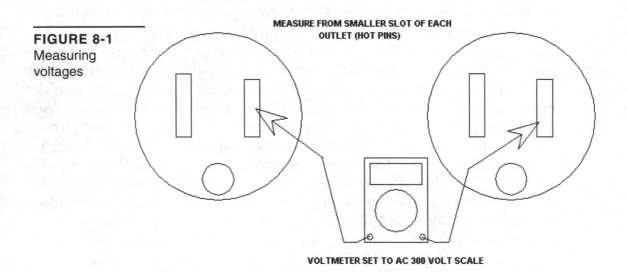

FIGURE 8-1
Measuring
voltages

MEASURE FROM SMALLER SLOT OF EACH
OUTLET (HOT PINS)

VOLTMETER SET TO AC 300 VOLT SCALE

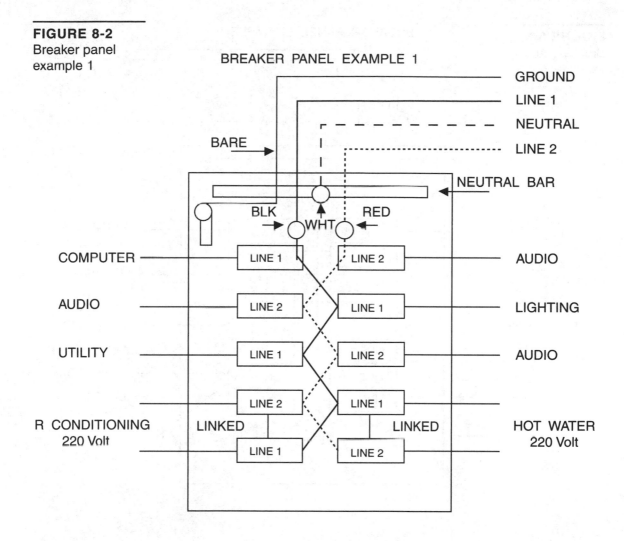

FIGURE 8-2
Breaker panel
example 1

    Note that all audio circuits are on one of the lines (2) exclusively, and
all other devices and lights are on the other line (1). This is to keep the
line where the audio equipment is connected separate from all the others. The zigzag lines are the internal connections in the box. Boxes are
done this way so that a pair of linked breakers, such as the two at the
bottom, can be used to protect a 220-volt feed together.

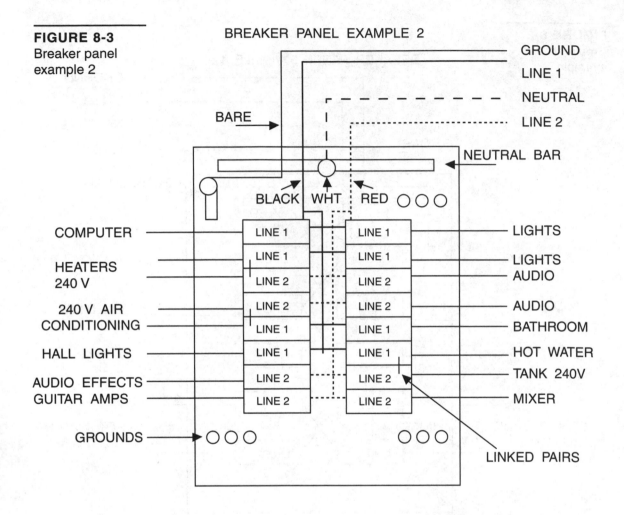

FIGURE 8-3
Breaker panel
example 2

This configuration ensures that noisy dimmers and appliances do not inject noise onto the same power phase as the audio power, causing buzzing sounds in the system. Also, always keep lighting on a different phase from the audio (see Figure 8-3).

FIGURE 8-4
An older panel
with fuses

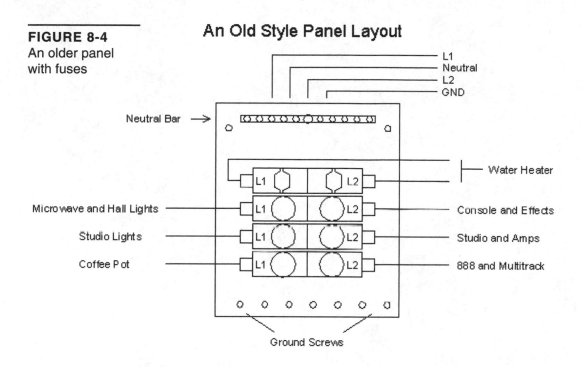

An Old Style Panel Layout

This is a different type or make of panel, and it may have a different pattern than another brand. Note that all audio circuits are still on one of the lines (2) exclusively, and all other devices and lights are on the other line (1). Figure 8-4 is an older style panel with fuses.

SIGNALS AND EQUIPMENT

CHAPTER NINE

Balanced and Unbalanced Signals

It is important to keep two types of analog audio signals in mind when connecting equipment together or when wiring a patch bay. These two audio signals are balanced and unbalanced (see Figures 9-1 and 9-2).

An unbalanced signal is a single conductor signal cable used for short-distance and semiprofessional applications. It is also used in consumer audio. Unbalanced cables have a tendency to pick up stray interference and hum if any current is allowed to flow in the shield or braid surrounding the core of the cable.

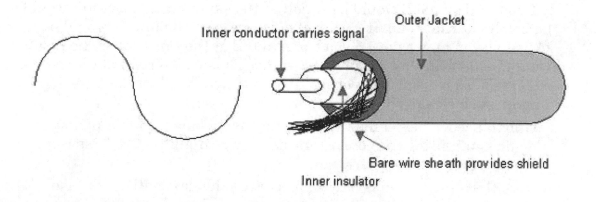

FIGURE 9-1
Unbalanced cable

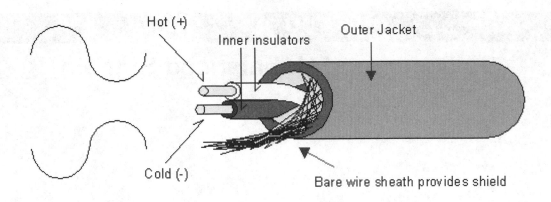

FIGURE 9-2
Balanced cable

A balanced signal has two waveforms opposite each other travelling in the two wires wound together inside the cable. This provides the advantage of being able to run this type of signal for long distances and pick up little noise. The noise immunity in this configuration is called *common mode rejection*. This means that any signal that appears on both wires in the cable will be rejected because the circuit receiving the signals only responds to a difference between the two signals, which would be the audio only (see Figure 9-3).

Try to think of the resulting signal C as A minus B, where A may be at +1 volt and signal B would be -1 volt at the same time. The result would be 2 volts at that time. If 3 volts of noise were on the line, it would be in phase, so you may have +3 volts on A and B at the same time. The result would be 0 because +3 minus +3 = 0. This is called *common mode rejection*. It is how long runs of signals are transmitted without noise getting in, such as early telephone networks.

Unbalanced cables can be improved by using balanced microphone cable and configuring the connectors, as shown in Figure 9-4. This configuration reduces triboelectric noise.

Note that the ground is only connected at the destination end and not at the source. This prevents a signal current from flowing through the ground sheath. The result is less noise getting on to the signal line inside.

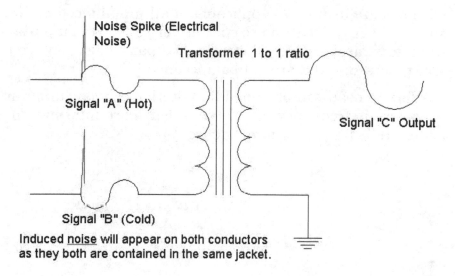

Noise Spike (Electrical Noise)

Transformer 1 to 1 ratio

Signal "A" (Hot)

Signal "C" Output

Signal "B" (Cold)

Induced <u>noise</u> will appear on both conductors as they both are contained in the same jacket.

FIGURE 9-3
Common mode rejection

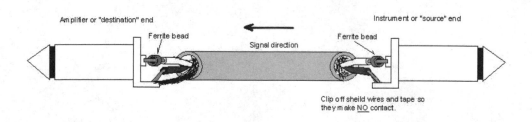

Amplifier or "destination" end

Ferrite bead

Signal direction

Instrument or "source" end

Ferrite bead

Clip off sheild wires and tape so they make <u>NO</u> contact.

FIGURE 9-4
Improving unbalanced cables by using balanced microphone cable

This configuration is applicable to all unbalanced audio connections such as RCA, $\frac{1}{4}$-inch, and so on. It is very effective for guitars and can be used for almost any unbalanced signal path, such as wiring to inexpensive processors that are not balanced.

A ferrite bead can be added for additional noise immunity. This bead blocks radio frequency interference. For more information, see chapter 33 on radio frequency interference.

CHAPTER TEN

Metering—VU, RMS, PPM and Correlation

Several types of measurements exist for voltages, decibels, current, and power. They are average, RMS, peak, and peak to peak. The most common is the average reading. This is what you would find on small consoles, DJ mixers, tape recorders, and amplifiers. This meter is also known as a *Volume Units* (VU) meter. The VU meter only reads the nominal levels in the unit and does not reading the actual output of the device in *decibels* (dB). In professional recording consoles, the VU meters read 0 with +4 dBu (dB undetermined impedance) at the output.

More elaborate consoles and some multitrack recorders have *peak program meters* (PPM) that may also be a type of VU meter, depending on what the 0 refers to. In a VU meter, the 0 means a unity gain where the output is at reference level. The fader 0 position means no gain or attenuation is present. What goes in comes out at the same level.

Compressors and other dynamics devices sometimes use RMS meters to give a true reading of RMS values for better energy reading accuracy. Table 10-1 is a chart for the conversions of each of these.

Meters can have different time constants or ballistics. Some may react very fast for rises and decays (average), whereas some may be rather slow (RMS). Others, such as PPM types, rise quickly and have a slow decay. These differences can be recognized with some experience.

Another type of meter is very useful in a console. It is called a correlation or phase meter. Its function is to show whether the two compared channels are in or out of phase for the majority of their audio program content. The meter locks on to the largest and lowest common frequency from each of the two audio channels and adds them together. It then

Table 10-1 Measurement conversion chart

From	Multipliers to get			
Given value	Average	RMS	Peak	Peak to peak
Average	——	1.11	1.57	3.14
RMS	0.9	——	1.414	2.828
Peak	0.637	0.707	——	2.0
Peak to peak	0.32	0.3535	0.5	——

displays them as positive for in phase and negative for out of phase. This type of meter is also useful for adjusting the tape recorder azimuth when connected to the console.

Most meters read from −20 to +5 dB, and for the most part these are Vu meters. PPM and RMS meters usually read from −40 to +5 and are far more accurate. Some correlation or phase meters read from −1 to +1 with 0 in the center. Some phase meters start at 0 and only read negative values. Figure 10-1 is a typical meter motor. Figures 10-2 to 10-4 are various meter scales and are identified as to the type of level displayed.

FIGURE 10-1
VU-style meter
motor

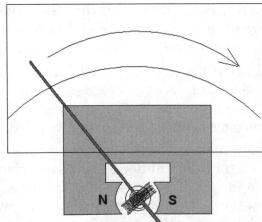

As the coil is energized, the lower end of the coil is energized with S which rotates the assembly.

N S

There is a return spring so that more energy is required to move it farther.

FIGURE 10-2
Peak Program
Meter Scale

FIGURE 10-3
VU-style meter
scale

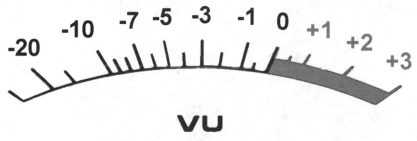

FIGURE 10-4
Peak or PPM
meter scale

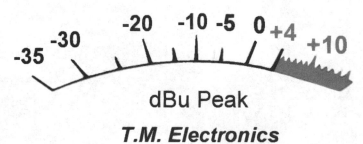

FIGURE 10-5
Correlator or
Phase meter
scale

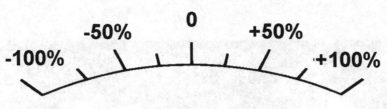

CORRELATION

CHAPTER ELEVEN

RF Transmissions—AM and FM

Several types of transmissions exist, but the most common in audio are AM and FM. AM means *amplitude modulation*. It is one of the oldest forms of broadcasting intelligible audio.

AM works with a transmitted carrier frequency. Let's use 550,000 Hz (550 kHz) as an example. To send audio, we need to create a modulation that can be mixed with the carrier. The easiest way is to vary the output power of the transmitter with the audio so that the output of the transmitter varies at the audio rate. This is the easiest form of transmission to detect. Figure 11-1 shows two waveforms are sine-wave-modulated test tones in *radio frequency* (RF)-modulated waveforms.

FIGURE 11-1
A sine wave amplitude modulated RF waveform

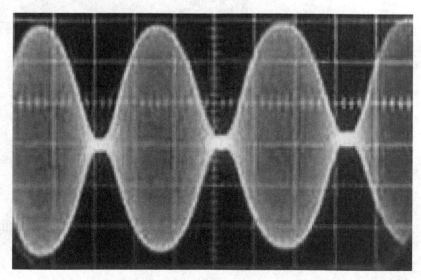

Figure 11-2 uses a lower-frequency carrier wave so it can be seen in detail. The envelope is at 10 kHz and the carrier is at 250 kHz.

Figure 11-3 is an audio signal modulating an RF carrier.

The audio portion of the waveform would look like the Figure 11-4.

The highlighted line is approximately what the audio content looked like to create the transmission envelopes. The higher, more positive peaks created a larger amount of power and the negative peaks reduced the power.

Frequency modulation (FM) transmission is a little more complicated. It uses a steady output power level and shifts the frequency of the transmitted signal with the audio frequency (see Figure 11-5).

The carrier frequency is centered at 100 MHz and is shifted to a higher frequency during the positive half-cycle of the audio signal. It is then shifted to a lower frequency during the negative half-cycle of the audio signal.

Because FM is not amplitude modulated, the receiver does not see amplitude interference like AM does. The amplitude-related noises are clipped off in the receiver, and the frequency shifts are decoded as audio.

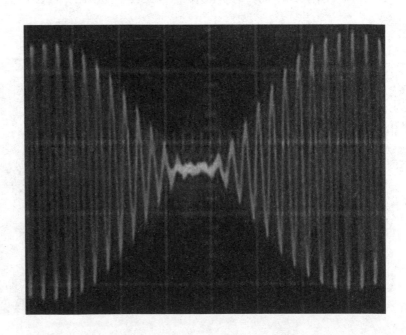

FIGURE 11-2
A lower frequency
carrier wave

FIGURE 11-3
Audio (music)
modulation

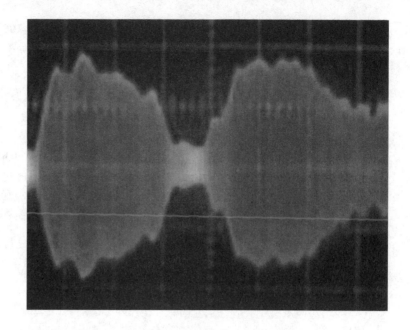

FIGURE 11-4
The waveform's audio
content

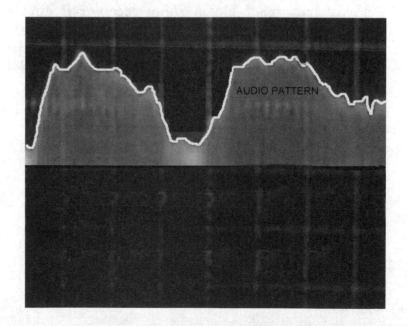

AUDIO PATTERN

FIGURE 11-5
FM transmission

FM RESULTANT WAVE   100 MHZ

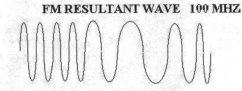

MODULATION WAVE   1 KHZ

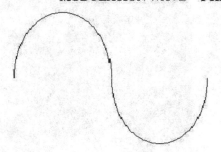

Interference

The most likely form of RF interference you will get in your audio gear is lower-band AM. The reason for this is that the front end of an audio device can react like a radio receiver if enough radiated RF signal can get in. FM does not usually get into the system because it is not amplitude modulated and the carrier frequency is too high. An AM radio receiver can be created easily and in a lot of cases by accident. Figure 11-6 shows how a simple AM radio receiver is made.

The diode is where the detection takes place. The circuitry in the pre-amp stages sometimes closely mimics this.

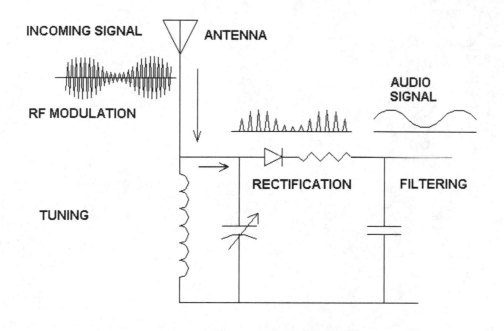

FIGURE 11-6
Creating an AM radio receiver

CHAPTER TWELVE

Compressors and Limiters

Compressors are devices that control the amount of dynamic range in the signal path. The device acts like a fast volume control to keep the loud parts from being too loud and the quiet parts from being too quiet.

This is accomplished by detecting the audio levels and converting them to a varying DC control signal. This signal is then fed to a variable attenuator or a *voltage-controlled amplifier* (VCA). The speed and response of this device is determined by adjustments according to the user's requirements. To track a vocal performer who is moving around and not focusing on the microphone, one may need to have a fast response and a rather severe amount of control to smooth out the variances in the level. If one is making the levels all relatively even in a mastering session for a CD, one would probably have a fast attack and a slow release in combination with a limiter to prevent overloading. This combination could assist in keeping the levels stable even while suppressing some of the higher peaks.

Figure 12-1 is a basic layout of what may be expected on the front panel of a compressor or limiter. Filters and options are not shown. Figure 12-2 is a basic layout of the front panel of an expander and gate. Filters and options are not shown.

The signal is introduced through the input-level circuit to the VCA, passed on to the buffer amplifier, and finally to the output. A sample of the output is fed to a switch where the user decides to use the internal or an external signal for the detector circuit. In the detector, the audio level is measured and the varying DC voltage is created from the changing volume of the audio. This voltage is adjusted and shaped by the timing circuits to vary the attack and release times. It is then applied to the link switch where it can be passed to the VCA.

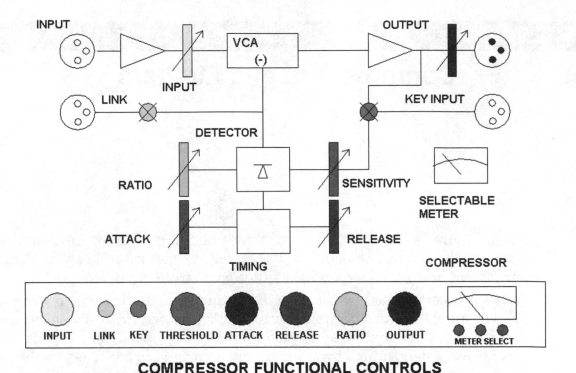

COMPRESSOR FUNCTIONAL CONTROLS

FIGURE 12-1
Basic compressor layout

If the user engages the link switch, two or more compressors can be tied together for stereo tracking. This stereo tracking would be necessary for mastering, because if the compressors were not linked, the audio triggering one compressor or limiter would cause that channel to drop slightly while the other would not. This effect would cause the stereo image to shift as though someone were moving a balance control back and forth. This is a most undesirable effect and is cured by having the VCAs track together while being triggered by either channel detector.

Many different types of compressors or limiters are available on the market and appear in many styles. Optical types use a light-sensitive resistor affected by a light source driven by the detector. *Current-controlled amplifier* (CCA) types require a slightly different type of detector. Vacuum

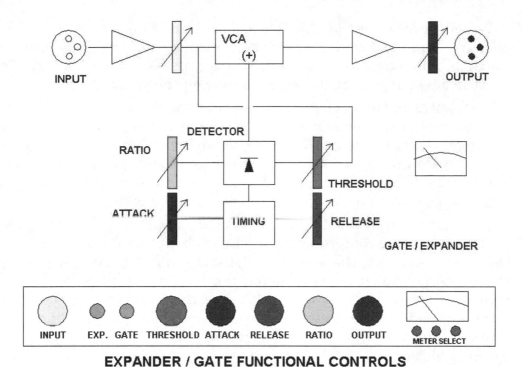

FIGURE 12-2
Expander and gate layout

tube types are still being used and have a unique sound. Also, *digital signal processor* (DSP) types can be used, which can stand alone or be integrated into digital mixing consoles and computer-based virtual mixers. Some manufacturers use a high-speed chopping circuit to slice up the audio and reassemble it while adjusting each sliced-up piece . Each and every style of compressor or limiter has its own advantage, disadvantage, and flavor of sound.

The earliest limiters date back to when AM radios were created and employed *automatic volume control* (AVC) circuits to keep the relative volume from changing despite how weak or strong the station was. AM radios change the gain of the *radio frequency* (RF) section of the radio receiver to keep the level stable. This process is always necessary for AM radios due to the nature of the transmitted signal.

Automatic level control (ALC) is a slow form of limiter and is used for portable, handheld digital and tape recorders. Reporters use the ALC function quite often while interviewing guests so that both the reporter and the guest appear at the same volume on the same recording.

Noise gates and expanders are the opposite of compressors and limiters. A noise gate keeps the signal attenuated until enough signal is present to trigger a user set value. When this value is reached, the gate begins to open and allow a signal through. In the case of an expander, this process is more linear and less drastic. The expander recovers the lost dynamic range of a signal that may have been compressed previously.

Some tape noise reduction systems operate this way. For example, to record, the system compresses at a ratio of about 2 to 1. During playback, the system expands the signal by the same amount. The thresholds of these systems must exactly match the record and play levels of the recorder they are working with; otherwise, the errors that occur will produce very undesirable results. These noise reduction systems had several brand names and were very effective only if the recorders were accurately calibrated.

CHAPTER THIRTEEN

Multitrack Recorders

General

This section examines all types of tape recorders, including digital ones. Small digital tape recorders usually do not require as many mechanism control circuits but require more audio- and digital-processing circuits. The heads are smaller but similar. The basic operation is nearly identical for all of them, from the small *digital audio tape* (DAT) machine all the way up to the large 2-inch, 24-track device and the 48-track dash machine.

Multitrack tape recorders are elaborate, powerful devices that can be very gentle and handle tape nearly perfectly if set up and maintained properly (see Figure 13-1). If they are poorly maintained or improperly adjusted, they can destroy tapes and remove human fingers very quickly. Understanding how a tape machine works is important for any kind of proper maintenance to take place. I strongly recommend reading the operating/service manual prior to *any* adjustments or disassembly.

Note that the circuit cards are arranged such that the transport control and *central processing units* (CPUs) are located on the top row. The lower three rows are all audio, arranged in groups of four cards to make one compete channel with eight channels per row.

FIGURE 13-1
A 24-track tape recorder

FIGURE 13-2
A 24-track drive system

FIGURE 13-3
$\frac{1}{4}$-inch drive and electronics

(A)Pinch Roller Solenoid
(B) Capstan Motor
(C) Capstan Speed Control
(D) Power Supply
(E) Line Amplifier
(F) Playback Amplifier
(G) Record Amplifier
(H) Bias and Erase Oscillator

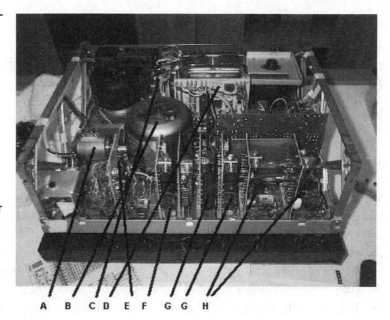

A   B   C D   E F   G G H

The Anatomy

1. The mechanism pulls the tape across the heads at a predetermined speed with as much accuracy as possible to minimize variations in speed and tension. This is accomplished by pinching the tape between an accurately controlled capstan and a pinch roller.

2. The heads transfer the information to and from the tape in a way that is compatible with as many other machines as possible. From left to right, when facing the machine, are the erase head, the record head, and the play head. These heads convert electrical energy into magnetic energy and place it precisely onto the tape. They recover the magnetic energy as well.

3. The audio electronics drive the audio-modulated electrical energy into the heads to retrieve the energy printed on the tape.

4. The mechanism electronics drive the motors and operate all the necessary brakes and levers to put the machine into various modes of operation.

The Mechanism

1. The basic mechanism consists of two reel motors, one capstan motor, one pinch roller, one erase head, one record head, one play head, various guides, and tape lifters.
2. The reel motors are usually directly connected to the reel tables and have a set of brakes that engages once the tape has been slowed down from fast wind to almost stopping.
3. The capstan motor is usually the rotating shaft that the pinch roller compresses the tape against to pull the tape at a specific speed (it moves at 7.5, 15, or 30 inches per second).
4. The erase head removes all recorded information from the specific track(s) that is selected for recording. The record head transfers the information from electrical energy into magnetic energy and focuses it onto the tape on a track where the erase head just cleaned magnetically. The play head retrieves the information from the tape after the recording process. The record head can also be used for retrieving the information in what is called *sync mode*, which is used for overdubs. See the following "Heads" section for more details.

The Electronics

The audio electronics accurately record and recover the audio information sent to the tape recorder. Many formats and levels are used for different standards. The machine can be calibrated to each of these formats and standards:

1. The components of the audio electronics would be playback amplifiers, sync amplifiers, record amplifiers, bias amplifiers, sync switching, line input amplifiers, line output amplifiers, master oscillators, meter amplifiers, and so on. The audio electronics are usually sets of circuit cards that perform three basic functions. The first of these is the bias and erase circuit, which cleans information from the tape in a track that is being recorded onto. It also generates a bias signal of about 150 kHz. The second is the record circuit, which sends the audio information mixed with the bias to the record head. The third is the play circuit, which recovers the information from the play head in repro mode or the record head in sync mode.

2. The mechanism electronics get somewhat complicated, but they can be simplified as well. The reason this section is complicated is that almost all the circuits must talk to the other circuits to know what the others are doing; otherwise, tape damage would be a common problem.

3. One of these circuits is the reel motor drive. It must drive the reel motors with specific amounts of current and voltage in different modes to pull the tape at particular amounts of force in all these modes (rewind, play, fast-forward, cue, edit, spool, and so on).

4. The next is the capstan control circuit, which must keep the capstan motor turning at a specific speed in order to pull the tape at a specific rate. The rest of the circuits drive the brakes, the pinch roller, the tape lifters, the tape counter, the motion and direction detection, and many fail-safe circuits to protect the tape from damage.

Heads

The tape heads consist of thinly sliced alloy metals of either permalloy or cobal alloy. These materials are usually amorphous (fast frozen from molten)

FIGURE 13-4
A complete 24-track head block assembly

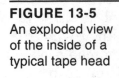

FIGURE 13-5
An exploded view
of the inside of a
typical tape head

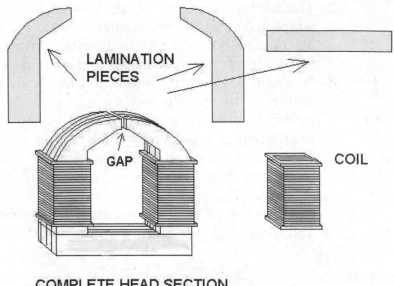

to reduce the crystallization in order to allow for thinner slicing. The more slices, the better the efficiency of the head. The other components are coils, spacers, terminals, shields, and the shell that contains the assembly.

The gap, located at the point of contact to the tape, creates a bulge in the magnetic flux that is made by the audio energy in the coils. This bulge gets printed onto the tape material (ferric oxide or iron rust) or is mixed with chromium dioxide. The magnetic pattern would look similar to the audio patterns on the edge of a movie film. Because the tape has a memory to it, the recording process has to have what is called bias mixed with the audio signal. This bias is usually a high-frequency signal about 100 times larger than the audio signal, depending on the frequency and equalization of the audio.

CHAPTER FOURTEEN

Speakers and Microphones

The speaker driver consists of several components: the frame, magnet, armature, voice coil, spider, cone, dust cap, surround, gasket, and tinsels (see Figure 14-1).

The way a speaker works is based on the left-hand rule of electromagnetism. Fingers point to the direction of current flow – thumb points to north created pole. If the current is reversed by reversing the voltage to the coil, the north pole switches to the other end of the coil (see Figure 14-2).

Now, let's insert a permanent magnet into the coil and see what happens when we apply a voltage, causing current to flow. The coil's right end becomes a north pole that repels the permanent magnet's north pole. At the

FIGURE 14-1
Speaker operation

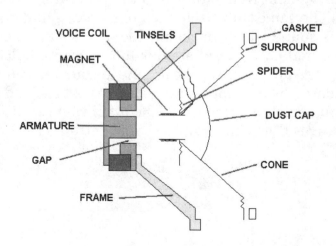

FIGURE 14-2
Polarisation of an
Electromagnet

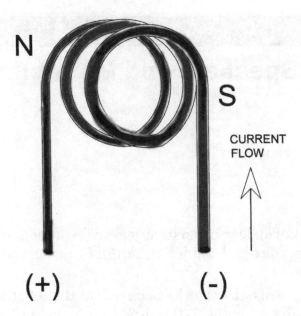

N

S

CURRENT
FLOW

(+)   (–)

same time, the coil's south pole is attracted to the permanent magnet's north pole. The result is that the coil moves to the right. Reversing the flow of current causes the coil to draw inward to the left (see Figure 14-3).

These are the reactions that make electric motors, generators, alternators, relays, solenoids, speakers, and microphones operate. A motor creates movement from power, and the generator creates power from movement. The same thing is applied to speakers and microphones. A speaker creates sound waves in the air like ripples in a pond. These waves' amplitude and pattern are similar to the original electric signal. You hear this by the air vibrating the organs in your ear. A microphone diaphragm is actuated by the ripples in the air like a boat tied to a dock. This moves the coil, which is attached to the diaphragm, which creates electric current waves that appear in the same patterns as the waves in the air.

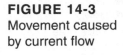

FIGURE 14-3
Movement caused
by current flow

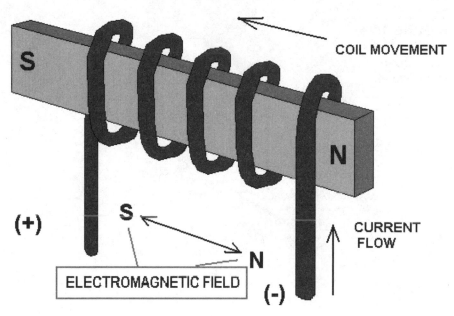

Carbon Microphones

The early microphones were made from compressing carbon in a chamber. The vibrations from one's voice made the carbon compress and expand with the sound waves. This crude microphone was used in the telephone industry where high fidelity was less important (see Figure 14-4).

One wire would have to be connected to the diaphragm, while the other is attached to the body, and both the body and diaphragm would be made of metal. A voltage would be applied to the microphone through a resistor that matched the microphone's resistance. As the sound waves varied the compression of the carbon, the resistance of the microphone would change, and as a result a varying voltage would be produced. Tapping on the microphone to see if it is on comes into play here. In some cases, the handset of the early telephones had to be tapped to get the carbon to loosen up enough to work.

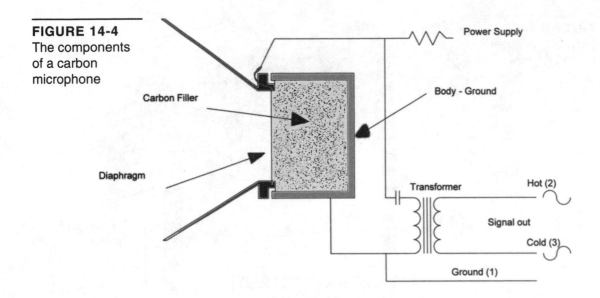

FIGURE 14-4
The components of a carbon microphone

Ribbon Microphones

The next generation of microphones were ribbon types. These microphones used a thin, flat piece of conductive material that was nonmagnetic. The material was suspended between the poles of a large, high-powered magnet. The current induced into the ribbon was fed to a transformer, which converted the current into a voltage. These microphones are still in use today for special sound characteristics (see Figure 14-5).

The main reason for creating this type of microphone was that no method existed for producing fine enough wire to be able to make a coil in the magnetic field. The only wire available would have been much too large to wind onto a paper form, such as today's dynamic microphones.

Dynamic Microphones

Dynamic microphones look like flat versions of speakers and have a similar structure. They act in a similar manner, but in reverse. The movement of the coil in the magnetic field induces a voltage in the coil (see Figures 14-6 through 14-10).

FIGURE 14-5
The inner workings of a ribbon microphone

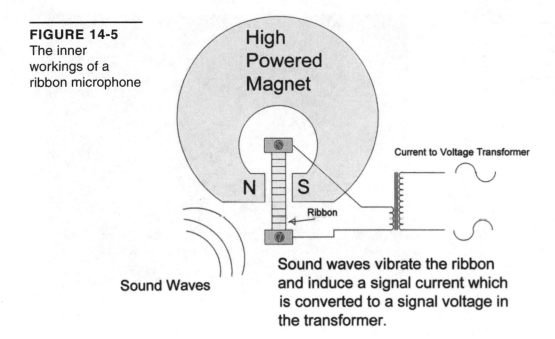

Sound waves vibrate the ribbon and induce a signal current which is converted to a signal voltage in the transformer.

FIGURE 14-6
Dynamic Microphone Diagram

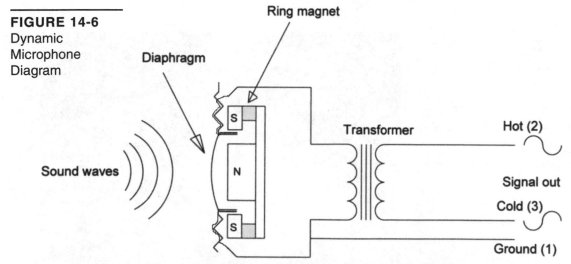

Sound waves vibrate the diaphragm resulting in a signal voltage which is sent through the transformer to the preamp.

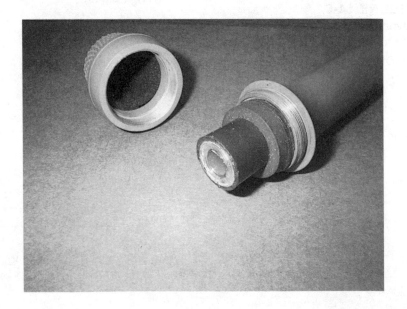

FIGURE 14-9
Dynamic
microphone
components

(A) Dome
(B) Coil
(C) Adhesive

FIGURE 14-10
A condenser
microphone

Air vibrations move the dome that moves the coil. This generates a small signal for the preamp.

Condenser Microphones

Condenser microphones are capacitors that store a charge on the diaphragm with respect to the backing plate (see Figure 14-11). As the sound wave presses the diaphragm inward, the capacitive action accumulates more charge and the voltage drops by a small amount through the resistor. As the sound wave reverses like ripples on a pond do, the diaphragm moves outward and can no longer store the previous amount of charge, and the voltage rises as the charge dissipates back through the resistor. This changing voltage is amplified and sent to the mixer or preamplifier (see Figure 14-12).

Self-Noise

A large diaphragm microphone has less self-noise than a small diaphragm one. This is because the self-noise in a microphone is mainly due to Brownian movements; that is, air molecules bombard the diaphragm,

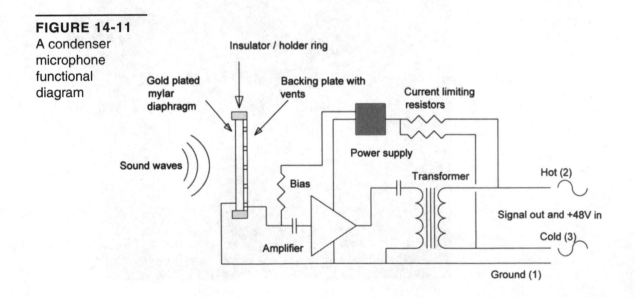

FIGURE 14-11
A condenser microphone functional diagram

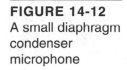

FIGURE 14-12
A small diaphragm
condenser
microphone

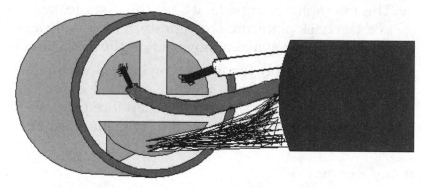

CONDENSER CAPSULES CAN HAVE
EITHER INTERNAL OR EXTERNAL BIAS
SHOWN HERE IS EXTERNAL BIAS (3 WIRES)

creating an equivalent noise pressure. The smaller diaphragm behaves like a hard surface, and the air molecules hitting it exchange a greater amount of their energy, producing greater sound pressure levels relative to the area and the sensitivity of the diaphragm.

Sensitivity

The sensitivity of the large and more compliant microphone diaphragm is generally higher than the small and stiff diaphragm. The large diaphragm is easier to move, even with low sound pressure levels, and therefore provides a larger output.

Sound Pressure Level (SPL) Handling

A condenser microphone's capability to handle large *Sound Pressure Levels* (SPLs) is limited by two things:

- The microphone capsule, where the distance between the diaphragm and the back plate and the rigidity of the diaphragm set a limit for how much a diaphragm can move before the distortion is too high.

- The power supply for the microphone preamplifier sets limits on the amount of signal that can be handled before clipping occurs. The smaller and stiffer diaphragm therefore can handle relatively higher SPLs than a microphone with a large diaphragm.

Frequency Range

As the omni-directional microphone senses small differences in the air pressure (sound waves), both large and small diaphragms are, in principle, equally capable of picking up low frequencies. The *lower limiting frequency* (LLF) of the pressure microphone is set by a small vent to prevent the diaphragm from moving due to changes in the ambient barometric pressure. According to the dimensions of the vent (diameter and length), it acts as an acoustic low-cut filter.

The *upper limiting frequency* (ULF) is set by several factors, all related to the dimensions of the diaphragm:

- A large diaphragm tends to break up and no longer acts as a true piston. This phenomenon is also known from loudspeaker technology and is the reason why loudspeakers are manufactured with different sizes of diaphragms to handle different frequency areas.

- The weight of the diaphragm attenuates the displacement of the diaphragm for higher frequencies.

- The diffractions around the edges of the microphone capsule limit the microphone's capability to handle very high frequencies. A large diaphragm microphone is considered to have a more limited frequency range than a small diaphragm.

Directional Characteristics

When a microphone is placed in a sound field, its mere presence influences the sound itself. This is because of the acoustic phenomena that can occur around the microphone due to the size of the microphone capsule, how it is positioned, the shape and the size of the microphone body containing the preamplifier, and the connector and design of the protecting grid. All flat-fronted omni-directional microphones become increasingly directional for higher frequencies. High-frequency sound waves coming directly from the front of the microphone will be reflected at the surface of the diaphragm, creating a sound pressure buildup between incoming and outgoing sound. This phenomenon occurs when the wavelength of the sound becomes comparable with or is smaller than the diameter of the diaphragm.

Dynamic Range

A small diaphragm microphone can usually offer a higher dynamic range than a large diaphragm mic. To explain this, it is useful to understand how the dynamic range is calculated.

The most sensible method of calculation is to establish the difference in dB between the noise floor and the SPL where the microphone produces a certain amount of *total harmonic distortion* (THD). We have seen earlier how the noise floor of the microphone rises if the diaphragm is small, but the SPL handling increases even more compared to the large diaphragm. Small diaphragm microphones can therefore have an equal or better dynamic range. The dynamic range is just shifted to cover different SPLs.

Figure 14-13 shows a small diaphragm condenser microphone (smaller than a dime) commonly used in telephones and portable recorders. These would also be found in podium microphones and in headset mics.

CHAPTER FIFTEEN

Wireless Microphones

Wireless Microphones

Wireless microphones and pickups are small-scale versions of FM and digital radio stations. They are also similar to cordless phones. They have a small transmitter mounted inside the body of the microphone or it is worn on the belt of the user. Typically, the use determines the type. Film mics use belt or pocket packs, whereas public speakers and DJs use handheld mics which have internal transmitters.

Unlike a sound wave, a radio wave includes both an electric field component and a magnetic field component (see Figure 15-1). The variations in these components have the same relative pattern along the direction

FIGURE 15-1
The electric and magnetic fields of a radio wave

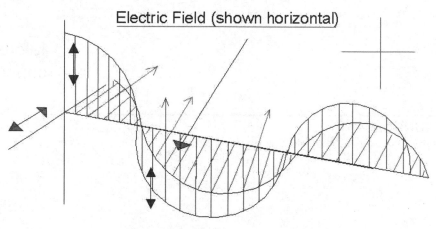

Electric Field (shown horizontal)

Magnetic Field (shown vertical)

of the radio wave, but they are oriented at a 90-degree angle to each other. In particular, the orientation of the electric field component determines the radio wave's angle of polarization. This becomes especially important when designing and adjusting antennas.

The sound waves from the speaker's voice are converted to electrical waves by the microphone. These electrical waves are amplified and used to modulate a RF signal or carrier wave created by the transmitter. This modulated RF signal is radiated from the transmitter by a small antenna that is either inside the body of the handheld microphone or extends from the belt transmitter.

The antennas of the receiver should be angled outward at the top and be clear of metal obstacles to capture all the possible positions of the transmitter while in use (see Figures 15-2 and 15-3).

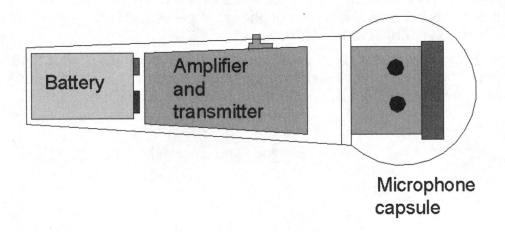

FIGURE 15-2
A typical DJ-type handheld mic transmitter unit

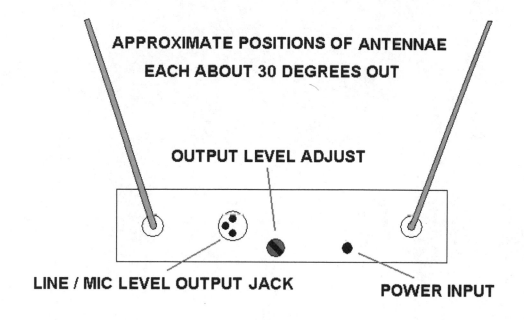

FIGURE 15-3
A typical DJ-type receiver unit

APPLICATIONS

Cables and Connectors

TABLE 16-1   Pin Connections for Connector Types

Audio	
XLR	Pin 2: hot or positive phase. Pin 3: cold or negative phase. Pin 1: *Ground* (GND).
$\frac{1}{4}$-inch	Tip: hot or positive phase. Ring: cold or negative phase. Sleeve: GND.
TRS	Tip: hot or positive phase. Ring: cold or negative phase. Sleeve: GND.
Bantam	Tip: hot or positive phase. Ring: cold or negative phase. Sleeve: GND.
RCA	Tip: hot or positive phase. Sleeve: GND.
DB25	Dependant on product manufacturer, this usually starts in reverse (high pins to lower channels).
Siemens 39	Pin 1 a: Ch. 1 (+). Pin 1 b: Ch. 1 (−). Pin 1 c: Ch. 1 Ground. Pin 2 a: Ch. 2 (+). Pin 2 b: Ch. 2 (−). Pin 2 c: Ch. 2 Ground.

Data		
DB9	Serial	Pin 1: Data Carrier Detect. Pin 2: Receive Data. Pin 3: Transmit Data. Pin 4: Data Terminal Ready. Pin 5: Data Signal GND. Pin 6: Data Set

TABLE 16-1 Pin Connections for Connector Types

DB25	Serial	Ready. Pin 7: Request to Send. Pin 8: Clear to Send. Pin 9: Ring Indicator. Pin 2: Transmit Data. Pin 3: Receive Data. Pin 4: Request to Send. Pin 5: Clear to Send. Pin 6: Data Set Ready. Pin 7: Data Signal GND. Pin 8: Data Carrier Detect. Pin 20: Data Terminal Ready. Pin 22: Ring Indicator.

Data

RJ45	Serial	Pin 1: Request to Send. Pin 2: Data Terminal Ready. Pin 3: Transmit Data. Pin 4: Data Signal GND. Pin 5: Ref GND. Pin 6: Receive Data. Pin 7: Data Set Ready. Pin 8: Clear to Send.
IBM Mini Din 8		Pin 1: Data Terminal Ready. Pin 2: Clear to Send. Pin 3: Transmit Data. Pin 4: Ground. Pin 5: Receive Data. Pin 6: Request to Send. Pin 7: Data Carrier Detect. Pin 8: Receive Clock.
Apple Mini Din 8		Pin 1: Output Handshake or Data Transmit Ready. Pin 2: Input Handshake or Mac Ext. Clock. Pin 3: Transmit Data ($-$). Pin 4: Ground. Pin 5: Receive Data ($-$). Pin 6: Transmit Data ($+$). Pin 8: Receive Data ($+$).

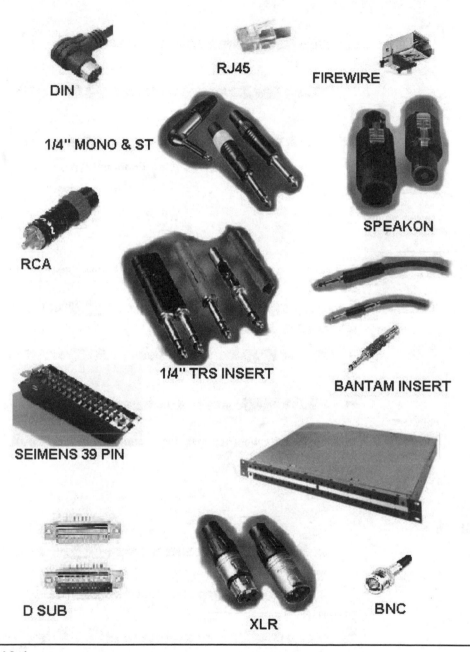

DIN

RJ45

FIREWIRE

1/4" MONO & ST

SPEAKON

RCA

1/4" TRS INSERT

BANTAM INSERT

SEIMENS 39 PIN

D SUB

XLR

BNC

FIGURE 16-1

FIGURE 16-2

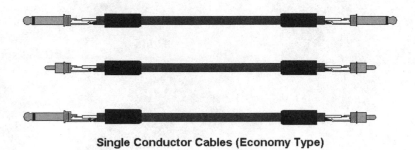

Single Conductor Cables (Economy Type)

FIGURE 16-3

From output of source device   To input of destination device

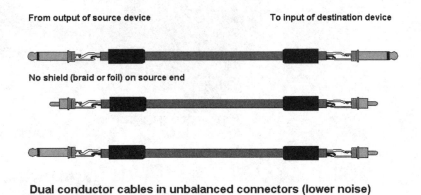

No shield (braid or foil) on source end

Dual conductor cables in unbalanced connectors (lower noise)

FIGURE 16-4

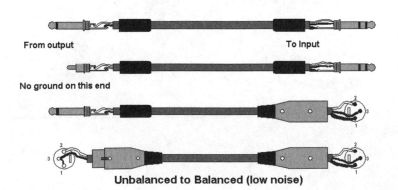

From output   To Input

No ground on this end

Unbalanced to Balanced (low noise)

FIGURE 16-5

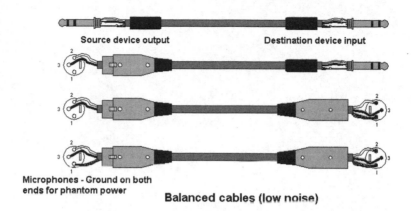

Source device output   Destination device input

Microphones - Ground on both
ends for phantom power

Balanced cables (low noise)

FIGURE 16-6
Connections to
Patch Bay Blocks

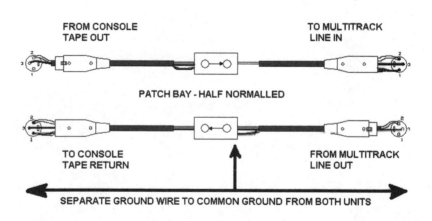

FROM CONSOLE
TAPE OUT   TO MULTITRACK
LINE IN

PATCH BAY - HALF NORMALLED

TO CONSOLE
TAPE RETURN   FROM MULTITRACK
LINE OUT

SEPARATE GROUND WIRE TO COMMON GROUND FROM BOTH UNITS

CHAPTER SEVENTEEN

Equipment and Rack Grounding

Equipment and Rack Grounding

Grounding the equipment in the studio can be done in many different ways. The most common style is called destination ground. This means that the grounds are only connected where the signal ends up. If you are connecting a mixer to a multitrack, you would connect the record outs to the tape recorder with the shields connected only at the tape recorder end. The tape returns to the console and would have the grounds connected only at the console end. Then the chassis of the console and recorder would be connected to a common ground bar under the floor or where the main studio ground for all the equipment is connected together. This would also be the case if a patch bay was in place. The ground would be connected at the destination of each termination. Figure 17-1 helps clarify this.

This wiring technique should remain consistent throughout the entire studio for every piece of gear. All chassis grounds should be brought to a central point in the studio, and the central point should be taken to earth ground. Microphones are an exception to this configuration, as they require grounds at both ends of their cables through to the console for phantom power.

FIGURE 17-1

RECORDER

PATCH BAY

EFFECTS RACK

MIXING CONSOLE

GROUND TO EARTH OR WATER PIPES

AMP RACK

FIGURE 17-2
Ground Routing Along Wall

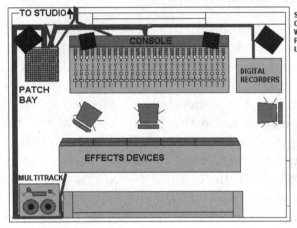

TO STUDIO

CONSOLE

DIGITAL RECORDERS

PATCH BAY

EFFECTS DEVICES

MULTITRACK

SHOWN IS A SAMPLE OF CABLE ROUTING WHERE IT IS NOT POSSIBLE TO RUN UNDER THE FLOOR.

CABLE BUNDLES SHOULD BE KEPT AS FAR AWAY FROM AC POWER CORDS AND AC OUTLETS AS POSSIBLE. WHERE THE TWO MUST CROSS, DO SO AT 90 DEGREE ANGLES TO AVOID INTERFERENCE.

FIGURE 17-3
Ground Routing Along Wall

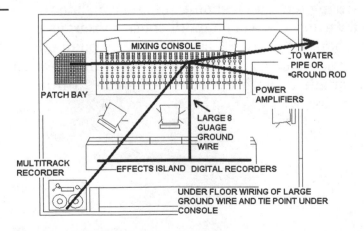

MIXING CONSOLE

TO WATER PIPE OR GROUND ROD

PATCH BAY

POWER AMPLIFIERS

LARGE 8 GUAGE GROUND WIRE

MULTITRACK RECORDER

EFFECTS ISLAND   DIGITAL RECORDERS

UNDER FLOOR WIRING OF LARGE GROUND WIRE AND TIE POINT UNDER CONSOLE

CHAPTER EIGHTEEN

Patch Bay Wiring

A patch bay is device with holes that allows you to change the connections by way of plugging alternate wiring patch cables into the holes. This activity causes internal switches to operate and thus re-route the signals. Patch bays are an important part of the studio when it comes to moving signal paths around to suit different applications. The patch bay may be a $\frac{1}{4}$-inch or a bantam style for most audio applications. The best way to understand how a patch bay works is to analyze what you need to do with the signal when you borrow from or move it.

Let's say you have a signal going from the console tape to the multitrack line. In order to change this, you need to pull the connection off the rear of the console and reconnect it. However, the cable connections are not designed to be moved around very often. The best way to get around this is to route the signal through a connector that redirects the signal when you push a jack into it. The symbols for the half-normal type are shown in Figure 18-1 (the most common type found). See Figures 18-8, 18-9, and 18-10

If you insert a plug into the left circle, you only borrow signal. The original signal will continue to flow to the right. If you insert a plug in the right side, you interrupt the former signal arriving, and your new signal via the plug is now introduced to the outgoing signal (see Figure 18-2).

If you insert a plug in either side, the original signal is broken and the new path is introduced (see Figure 18-3).

Let's look inside the connector (see Figure 18-4).

The tip and tip normal are connected together until a plug is inserted into the jack. The same setup is used for the ring and ring normal. When

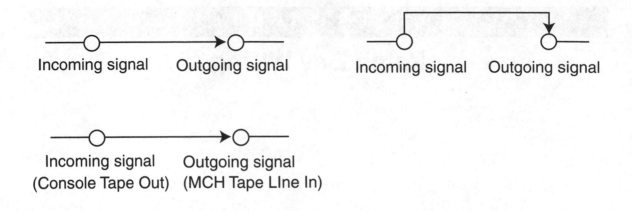

FIGURE 18-1
Half-normal type symbols (2 styles)

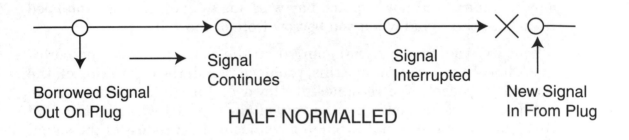

FIGURE 18-2
Half Normalled Type

the plug is inserted, the connections between the ring and ring normal and the tip and tip normal are broken by the lifting action of the plug displacing the contacts. If you want the new signal to replace the old one, the original signal must be connected to the ring normal and the tip normal (see Figure 18-5).

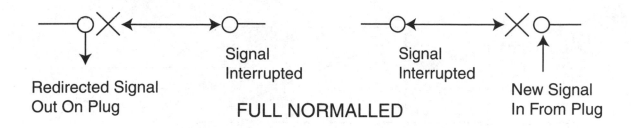

Redirected Signal
Out On Plug

Signal
Interrupted

FULL NORMALLED

Signal
Interrupted

New Signal
In From Plug

FIGURE 18-3
Full Normalled Type (microphones)

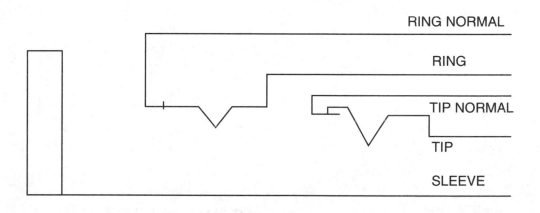

RING NORMAL

RING

TIP NORMAL

TIP

SLEEVE

FIGURE 18-4
Patch bay jack at rest (empty)

A typical example of this would be going from the console's tape to the multitrack recorder's line input. Let's say you want to insert a tone generator to calibrate the tape recorder. You would need to interrupt the original connection from the console and replace it with the generator connection. This type of connection is a *normalized* one. The mate to this

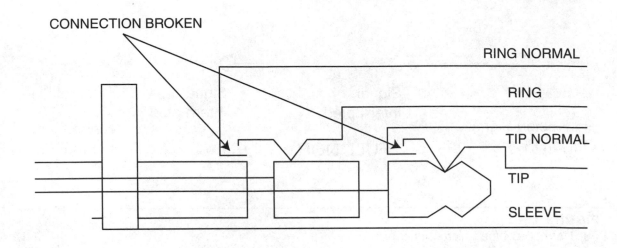

CONNECTION BROKEN

RING NORMAL

RING

TIP NORMAL

TIP

SLEEVE

FIGURE 18-5
An insert connector with the plug in

FIGURE 18-6
Front-view
representation

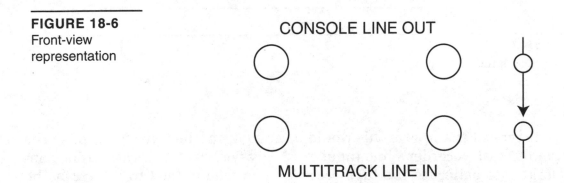

CONSOLE LINE OUT

MULTITRACK LINE IN

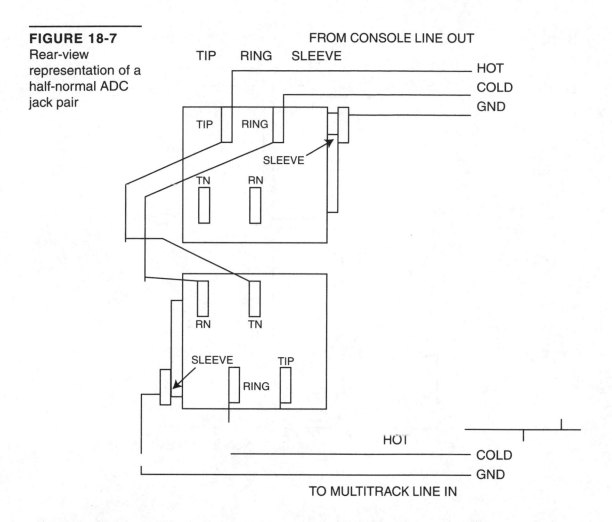

FIGURE 18-7
Rear-view representation of a half-normal ADC jack pair

FROM CONSOLE LINE OUT

TO MULTITRACK LINE IN

connection is the console out, and it would usually be located just above the recorder in (see Figures 18-6 and 18-7).

The simplest way to remember that the configurations are six wires on five pins is that it is a breaking connection. Six wires on three pins are a non-breaking connection because they are soldered together.

Also, remember to connect grounds at both ends of the microphone lines for row A and B to carry phantom power. Also wire the microphones in a full-normal configuration (see Figure 18-8). See the RF Interference

FIGURE 18-8
Rear view of a full
normalled
connection

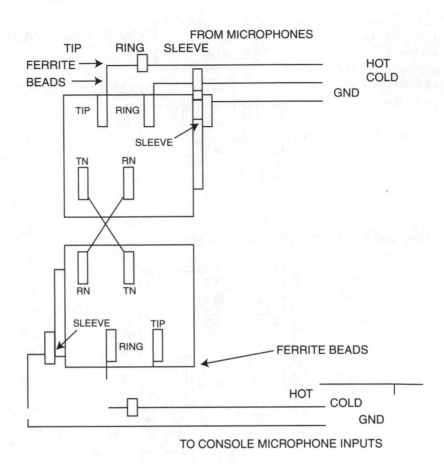

Chapter 33 of this book for details. Always remember *not* to install ferrite beads on the grounds.

The easiest way to lay out a patch bay is to look at the signal flow through the block diagram of the console. Start with the inputs to the channel strips and end at the monitoring section.

Figure 18-9a is a functional block diagram of an input channel of a mixing console showing the inserts and the tape sends and returns.

Figure 18-9 is a basic layout of one channel, one auxiliary, one two-mix, and so on. The effects and other peripheral devices are not shown. Almost all the ins and outs of the console are connected through the patch bay, which is where some of the grounding is configured as well.

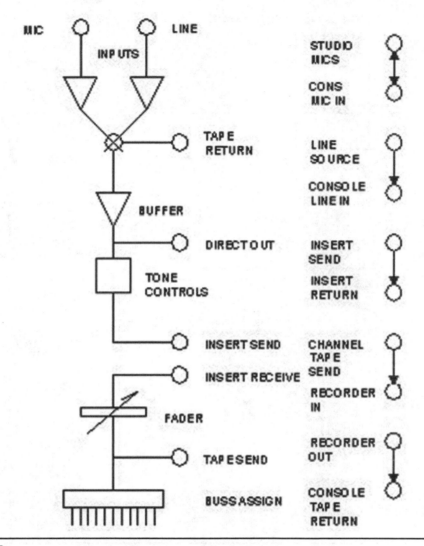

FIGURE 18-9a
A block diagram of a mixing console input module

Let's look at a typical patch bay configuration that may be a general purpose, 24-track application (see Figure 18-10). When designing your patch bay, you should always make a wire list so that all the paths are documented and traceable to their destinations. Table 2-1 shows a 16-channel example.

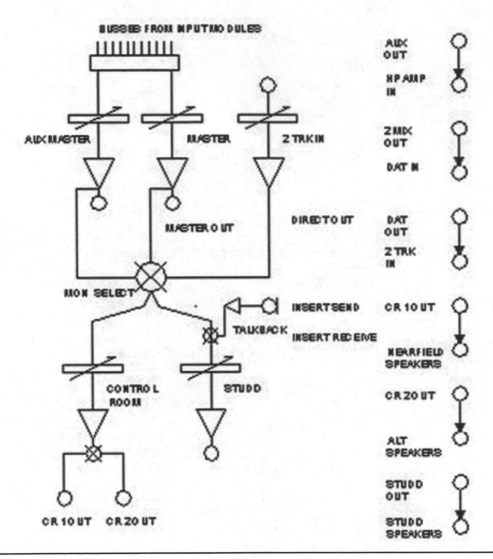

FIGURE 18-9b (CONT.)
Block diagram of a mixing console output module

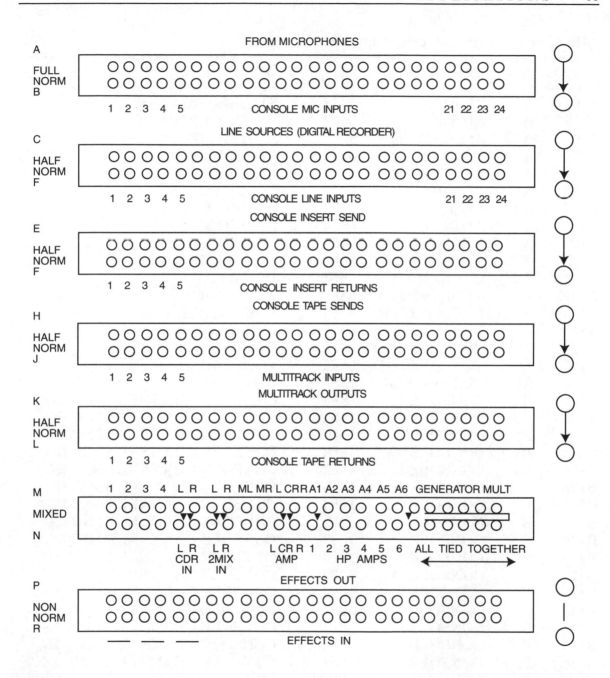

FIGURE 18-10
A typical small patch bay

TABLE 18-1 Sixteen-channel example

Wire no.	Row	Destination	Length	Description
A101	A	STUDIO	50FT	STUDIO MIC 1
A102	A	STUDIO	50FT	STUDIO MIC 2
A103	A	STUDIO	50FT	STUDIO MIC 3
A104	A	STUDIO	50FT	STUDIO MIC 4
A105	A	STUDIO	50FT	STUDIO MIC 5
A106	A	STUDIO	50FT	STUDIO MIC 6
A107	A	STUDIO	50FT	STUDIO MIC 7
A108	A	STUDIO	50FT	STUDIO MIC 8
A109	A	STUDIO	50FT	STUDIO MIC 9
A110	A	STUDIO	50FT	STUDIO MIC 10
A111	A	DRUM BOOTH	65FT	DRUM MIC 1
A112	A	DRUM BOOTH	65FT	DRUM MIC 2
A113	A	DRUM BOOTH	65FT	DRUM MIC 3
A113	A	DRUM BOOTH	65FT	DRUM MIC 3
A114	A	VOCAL BOOTH	55FT	VOCAL MIC 1
A115	A	VOCAL BOOTH	55FT	VOCAL MIC 2
A116	A	PIANO ROOM	75FT	PIANO MIC
B101	B	CONSOLE	18FT	CONSOLE CHANNEL 1 MIC IN
B102	B	CONSOLE	18FT	CONSOLE CHANNEL 2 MIC IN
B103	B	CONSOLE	18FT	CONSOLE CHANNEL 3 MIC IN
B104	B	CONSOLE	18FT	CONSOLE CHANNEL 4 MIC IN
B105	B	CONSOLE	18FT	CONSOLE CHANNEL 5 MIC IN
B106	B	CONSOLE	18FT	CONSOLE CHANNEL 6 MIC IN
B107	B	CONSOLE	18FT	CONSOLE CHANNEL 7 MIC IN
B108	B	CONSOLE	18FT	CONSOLE CHANNEL 8 MIC IN
B109	B	CONSOLE	18FT	CONSOLE CHANNEL 9 MIC IN
B110	B	CONSOLE	18FT	CONSOLE CHANNEL 10 MIC IN
B111	B	CONSOLE	18FT	CONSOLE CHANNEL 11 MIC IN
B112	B	CONSOLE	18FT	CONSOLE CHANNEL 12 MIC IN
B113	B	CONSOLE	18FT	CONSOLE CHANNEL 13 MIC IN
B114	B	CONSOLE	18FT	CONSOLE CHANNEL 14 MIC IN
B115	B	CONSOLE	18FT	CONSOLE CHANNEL 15 MIC IN
B116	B	CONSOLE	18FT	CONSOLE CHANNEL 16 MIC IN
C101	C	PRO TOOLS 888	10FT	PRO TOOLS 888 CHANNEL 1 OUT
C102	C	PRO TOOLS 888	10FT	PRO TOOLS 888 CHANNEL 2 OUT

TABLE 18-1   Sixteen-channel example (cont.)

Wire no.	Row	Destination	Length	Description
C103	C	PRO TOOLS 888	10FT	PRO TOOLS 888 CHANNEL 3 OUT
C104	C	PRO TOOLS 888	10FT	PRO TOOLS 888 CHANNEL 4 OUT
C105	C	PRO TOOLS 888	10FT	PRO TOOLS 888 CHANNEL 5 OUT
C106	C	PRO TOOLS 888	10FT	PRO TOOLS 888 CHANNEL 6 OUT
C107	C	PRO TOOLS 888	10FT	PRO TOOLS 888 CHANNEL 7 OUT
C108	C	PRO TOOLS 888	10FT	PRO TOOLS 888 CHANNEL 8 OUT
C109	C	PRO TOOLS 888	10FT	PRO TOOLS 888 CHANNEL 9 OUT
C110	C	PRO TOOLS 888	10FT	PRO TOOLS 888 CHANNEL 10 OUT
C111	C	PRO TOOLS 888	10FT	PRO TOOLS 888 CHANNEL 11 OUT
C112	C	PRO TOOLS 888	10FT	PRO TOOLS 888 CHANNEL 12 OUT
C113	C	PRO TOOLS 888	10FT	PRO TOOLS 888 CHANNEL 13 OUT
C114	C	PRO TOOLS 888	10FT	PRO TOOLS 888 CHANNEL 14 OUT
C115	C	PRO TOOLS 888	10FT	PRO TOOLS 888 CHANNEL 15 OUT
C116	C	PRO TOOLS 888	10FT	PRO TOOLS 888 CHANNEL 16 OUT
D101	D	CONSOLE	18FT	CONSOLE CHANNEL 1 LINE IN
D102	D	CONSOLE	18FT	CONSOLE CHANNEL 2 LINE IN
D103	D	CONSOLE	18FT	CONSOLE CHANNEL 3 LINE IN
D104	D	CONSOLE	18FT	CONSOLE CHANNEL 4 LINE IN
D105	D	CONSOLE	18FT	CONSOLE CHANNEL 5 LINE IN
D106	D	CONSOLE	18FT	CONSOLE CHANNEL 6 LINE IN
D107	D	CONSOLE	18FT	CONSOLE CHANNEL 7 LINE IN
D108	D	CONSOLE	18FT	CONSOLE CHANNEL 8 LINE IN
D109	D	CONSOLE	18FT	CONSOLE CHANNEL 9 LINE IN
D110	D	CONSOLE	18FT	CONSOLE CHANNEL 10 LINE IN
D111	D	CONSOLE	18FT	CONSOLE CHANNEL 11 LINE IN
D112	D	CONSOLE	18FT	CONSOLE CHANNEL 12 LINE IN
D113	D	CONSOLE	18FT	CONSOLE CHANNEL 13 LINE IN
D114	D	CONSOLE	18FT	CONSOLE CHANNEL 14 LINE IN
D115	D	CONSOLE	18FT	CONSOLE CHANNEL 15 LINE IN
D116	D	CONSOLE	18FT	CONSOLE CHANNEL 16 LINE IN
E101	E	CONSOLE	18FT	CONSOLE INSERT SEND 1
E102	E	CONSOLE	18FT	CONSOLE INSERT SEND 2
E103	E	CONSOLE	18FT	CONSOLE INSERT SEND 3
E104	E	CONSOLE	18FT	CONSOLE INSERT SEND 4
E105	E	CONSOLE	18FT	CONSOLE INSERT SEND 5
E106	E	CONSOLE	18FT	CONSOLE INSERT SEND 6
E107	E	CONSOLE	18FT	CONSOLE INSERT SEND 7

TABLE 18-1   Sixteen-channel example (cont.)

Wire no.	Row	Destination	Length	Description
E108	E	CONSOLE	18FT	CONSOLE INSERT SEND 8
E109	E	CONSOLE	18FT	CONSOLE INSERT SEND 9
E110	E	CONSOLE	18FT	CONSOLE INSERT SEND 10
E111	E	CONSOLE	18FT	CONSOLE INSERT SEND 11
E112	E	CONSOLE	18FT	CONSOLE INSERT SEND 12
E113	E	CONSOLE	18FT	CONSOLE INSERT SEND 13
E114	E	CONSOLE	18FT	CONSOLE INSERT SEND 14
E115	E	CONSOLE	18FT	CONSOLE INSERT SEND 15
E116	E	CONSOLE	18FT	CONSOLE INSERT SEND 16
F101	E	CONSOLE	18FT	CONSOLE INSERT RETURN 1
F102	E	CONSOLE	18FT	CONSOLE INSERT RETURN 2
F103	E	CONSOLE	18FT	CONSOLE INSERT RETURN 3
F104	E	CONSOLE	18FT	CONSOLE INSERT RETURN 4
F105	E	CONSOLE	18FT	CONSOLE INSERT RETURN 5
F106	E	CONSOLE	18FT	CONSOLE INSERT RETURN 6
F107	E	CONSOLE	18FT	CONSOLE INSERT RETURN 7
F108	E	CONSOLE	18FT	CONSOLE INSERT RETURN 8
F109	E	CONSOLE	18FT	CONSOLE INSERT RETURN 9
F110	E	CONSOLE	18FT	CONSOLE INSERT RETURN 10
F111	E	CONSOLE	18FT	CONSOLE INSERT RETURN 11
F112	E	CONSOLE	18FT	CONSOLE INSERT RETURN 12
F113	E	CONSOLE	18FT	CONSOLE INSERT RETURN 13
F114	E	CONSOLE	18FT	CONSOLE INSERT RETURN 14
F115	E	CONSOLE	18FT	CONSOLE INSERT RETURN 15
F116	E	CONSOLE	18FT	CONSOLE INSERT RETURN 16
H101	H	CONSOLE	18FT	CONSOLE TAPE SEND 1
H102	H	CONSOLE	18FT	CONSOLE TAPE SEND 2
	AND SO ON		TO	CHANNEL 16
H116	H	CONSOLE	18FT	CONSOLE TAPE SEND 16
J101	J	MULTITRACK	20FT	MULTITRACK LINE IN 1
	AND SO ON		TO	CHANNEL 16
K101	K	MULTITRACK	20FT	MULTITRACK LINE OUT 1
	AND SO ON		TO	CHANNEL 16
L101	L	CONSOLE	18FT	CONSOLE TAPE RETURN 1
	AND SO ON		TO	CHANNEL 16

At this point, you see the pattern emerging from the layouts of the console and the equipment lists. The block diagram and rear panel layout of the console and other gear will dictate what your patch bay will look like. Bear in mind that you may need to install ferrite beads on the patch bay and in the connectors if you are near an AM transmitter to minimize RF interference. See the RF Interference section of this book for details, and always remember *not* to install ferrite beads on the grounds.

CHAPTER NINETEEN

Cable Routing

Figures 19-1 through 19-6 show a proper way to dress and prepare cables in a patch bay and at the rear of a console.

Under Floor

The wiring under the floor should have the same considerations as the other cables. The audio and video should be kept as far away from the electrical cabes as possible, and where they need to cross, they should do so at right angles. Also leave enough wire looped to be able to pull the equipment out of the front while still connected. This allows for easy changes.

Figure 19-8 shows the routing in a control room where no subfloor is used to route signals, whereas Figure 19-9 shows a subfloor being utilized. Note the amount of wire-length reduction by routing the wires under the floor. Typically, access plates are used at certain intervals where turns and junctions occur.

FIGURE 19-1
Audio rack (view
from rear)

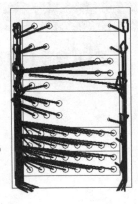

SIGNAL WIRES ALL
ROUTE DOWN ONE
SIDE WITH ABOUT 6
INCHES EXTRA
CABLE LOOPED SO
THAT EQUIPMENT
CAN BE EXTRACTED
FROM THE FRONT
WHILE REMAINING
CONNECTED.

POWER WIRES ROUTE
DOWN OPPOSITE SIDE OF
RACK AND ARE KEPT AS
FAR AWAY FROM SIGNAL
AS POSSIBLE. WHEN THE
TWO TYPES MUST
CROSS, DO SO AT 90
DEGREE ANGLES FOR
MINIMUM INTERFERENCE.

FIGURE 19-2
Proper way to
dress cables in a
patch bay

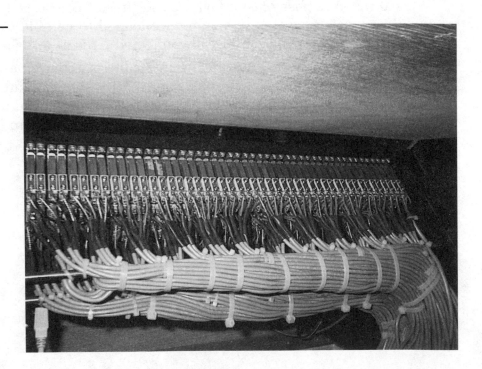

FIGURE 19-3
Different angle
view showing
punch block at
bottom

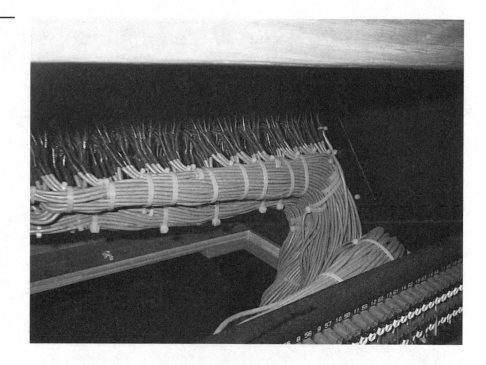

FIGURE 19-4
Allow for stress
relief on cables—
note folds in wires

FIGURE 19-5
Proper routing in
desk

FIGURE 19-6
Tieing up cables

FIGURE 19-7
Cable suspension

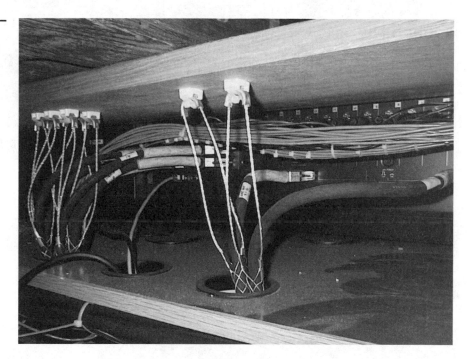

FIGURE 19-8

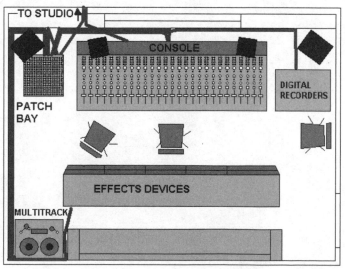

TO STUDIO

CONSOLE

PATCH BAY

DIGITAL RECORDERS

EFFECTS DEVICES

MULTITRACK

SHOWN IS A SAMPLE OF CABLE ROUTING WHERE IT IS NOT POSSIBLE TO RUN UNDER THE FLOOR.

CABLE BUNDLES SHOULD BE KEPT AS FAR AWAY FROM AC POWER CORDS AND AC OUTLETS AS POSSIBLE. WHERE THE TWO MUST CROSS, DO SO AT 90 DEGREE ANGLES TO AVOID INTERFERENCE.

FIGURE 19-9

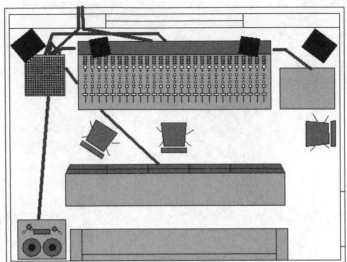

SHOWN IS AN UNDER-FLOOR ROUTING PATTERN WHICH SAVES MATERIAL AND KEEPS THE SIGNAL WIRES AWAY FROM WALLS WHERE AC POWER IS SOMETIMES ROUTED.

CHAPTER TWENTY

Tools

Tools

This chapter outlines some of the tools and materials to have on hand in the studio. The following list is suggested so you can handle most of the basic wiring and maintenance chores. More extensive procedures would, in most cases, require specific tools.

- **Soldering station**   40-watt or more, temperature controlled
- **Solder**   One millimeter, rosin core, $^{60}/_{40}$
- **Cutters**   Flush cutting with a spring opening
- **Pliers**   Sharp-nosed, 3-inch, or blunt nosed for removing nuts from controls
- **Wire strippers**   Adjustable or multisize with a spring opening
- **Vise**   Small, workbench style
- **Screwdrivers**   Philips, Robertson, Hex, Torx, and tiny, flat-blade sizes
- **Voltmeter/ohmmeter**   A basic type, with digital, continuity, and auto-ranging capabilities if desired
- **Spare connectors**   XLR, ¼-inch stereo, ¼-inch mono, TRS or Bantam, RCA, and so on
- **Extra cable**   Microphone (balanced), balanced hookup (Delco or equivalent), and heavy ground wire

- **Extra mounting hardware** Rack mount screws with washers (10–32)
- **Plastic parts bins** For storing extra screws and small valuable items
- **Flashlight** Maglight or equivalent small flashlight for inside racks and equipment
- **Magnifying lens** For examining heads and small mechanisms
- **Towels** For spreading over the console and the machine surfaces when maintenance is needed

CHAPTER TWENTY-ONE

Test Equipment

This chapter examines some of the test equipment to have on hand in the studio. This is an area that can get expensive, so evaluations are required to assess the real need for some of these items. This is a partial list, and as time goes on, one will get a better idea of what is necessary for the studio.

- **Cable checker** This tool has connectors on it so that cables can be plugged into it and tested for continuity.
- **Correlator or phase meter** For checking the phase relationship of one signal to another. It can also be used to verify cables.
- **Portable tone generator** For inserting signals into paths to test equipment.
- **Oscilloscope** For viewing waveforms of signals that may be distorted or have noise on them.
- **Transformer signal isolator** For breaking ground connections but allowing a signal to pass through. This is good for finding ground problems.
- **Portable powered speaker** For moving around and searching for signals.
- **Phase-inverting cables** For flipping the phase of a signal to correct a problem. A phase-inverting patch cord is a valuable item.
- **Phase-inverting adapters** An alternate to the previous item.
- **Active DI box** For correcting impedance problems, flipping phase, and so on.

- **Ground fault readers** Special AC plug with lights to signal power faults.

- **Test tapes** For aligning analog recorders.

- **Service manuals** For the larger items like the mixer and multitrack recorder. Service manuals are so important that I would suggest not buying a machine or console without a manual. This may cost a little extra, but it is essential later when service is required. I feel so strongly about this that I normally do not buy a piece of equipment without one, and that includes vehicles!

The Digital Volt/Ohm Meter (DVM)

This is a typical, basic meter for the service technician. It has several uses and functions. In some cases, the more expensive and complicated ones can create more confusion than assistance, but they can be purchased for under $100 (see Figure 21-1)

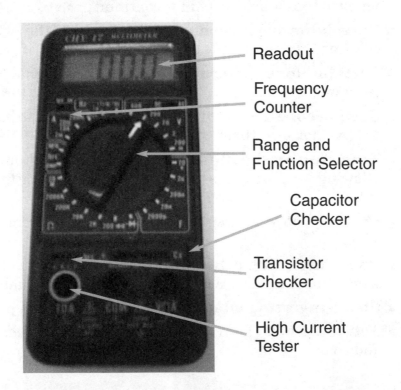

FIGURE 21-1
The Digital
Multimeter

DC Volts
AC Volts
DC Amps
AC Amps
Resistance
Capacitance
Frequency
Counter
Transistor Gain
Diode Tester
Continuity

Readout

Frequency
Counter

Range and
Function Selector

Capacitor
Checker

Transistor
Checker

High Current
Tester

The Oscilloscope

This is an economy type oscilloscope that will be adequate for almost all audio troubleshooting. Shown in Figures 21-2 through 21-4 on the display are Audio sine waves.

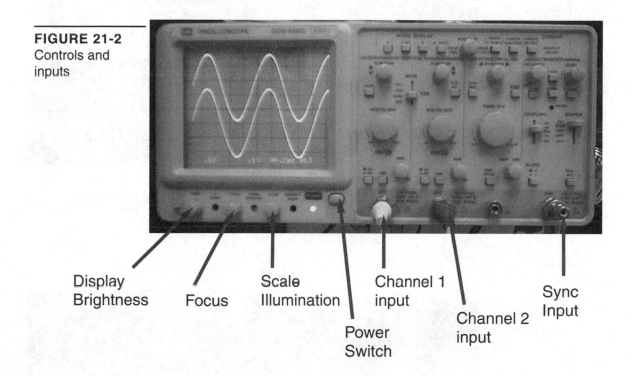

FIGURE 21-2
Controls and inputs

Display Brightness

Focus

Scale Illumination

Power Switch

Channel 1 input

Channel 2 input

Sync Input

FIGURE 21-3
Common controls

Sweep
Rate

Trigger
Source

Trace Positions

Input Gain/
Attenuators

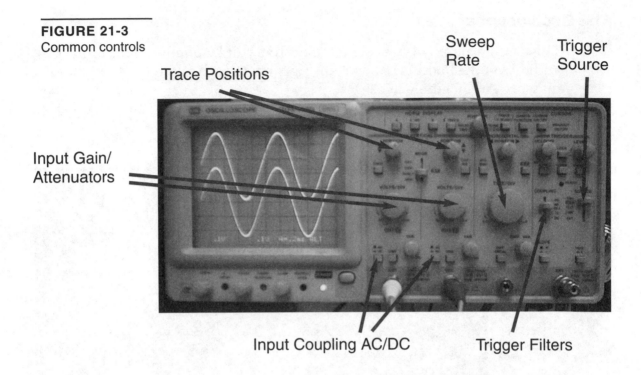

Input Coupling AC/DC

Trigger Filters

FIGURE 21-4

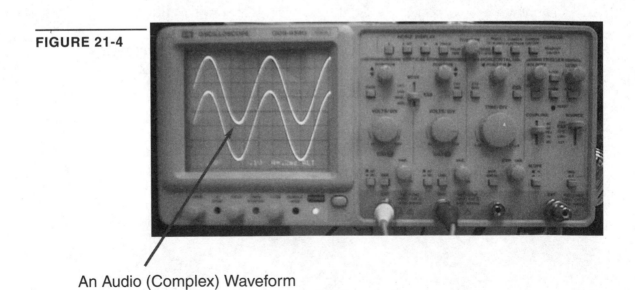

An Audio (Complex) Waveform

A 48-Volt Phantom Power Supply Project

In this chapter, you'll add a phantom power supply to your patch bay or studio microphone wall plate. This would be useful if phantom power is not available from your console.

The items needed are shown in Table 22-1.

TABLE 22-1   Materials List

Qty.	Designations	Part Number	Description
2	C1, C2 1000uF/63V		1000 microfarad, 63-volt electrolytic radial lead capacitor
1	C3	10uF/50V	10 microfarad, 50-volt electrolytic radial lead capacitor
2	D1, D2	1N4007	1000-volt, 1-amp rectifier diode
1	Q1	2SD880 or equivalent	Transistor, NPN, 3 amp
1	R1	1.5K ½ Watt	1500 ohm, ½-watt resistor (brown, green, red, gold)
1	R2	10R ½ watt	10 ohm, ½-watt resistor (brown, black, black, gold)
1	R3	5.6K ½ watt	5600 ohm, ½-watt resistor (optional) (green, blue, red, gold)
2	D3, D4	1N474x	24-volt, ½-watt Zener diode
1	D5	LED	5-milimeter green or red *light-emitting diode* (LED) (optional)
1			3-inch × 2-inch piece of aluminum sheet about ⅛-inch thick
1			3-inch perforated circuit board
1			4-inch × 5-inch **plastic box**

Construction

1. Drill a $\frac{3}{16}$-inch hole into each end of the plastic box for the power wire from the adapter.
2. Mount the components listed previously in the configuration shown in Figure 22-2.
3. Wire the underside of the circuit board to connect the components as shown Figure 2-7.
4. Double- and triple-check your connections so that you make no mistakes.

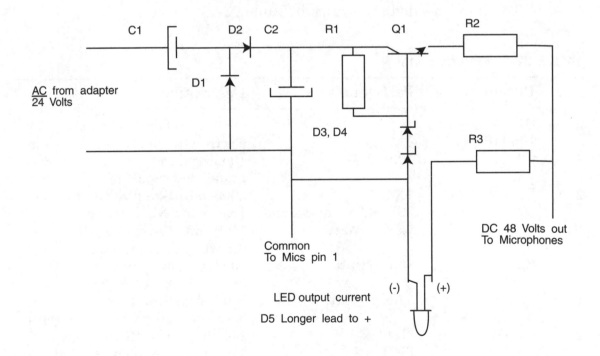

FIGURE 22-1
The schematic

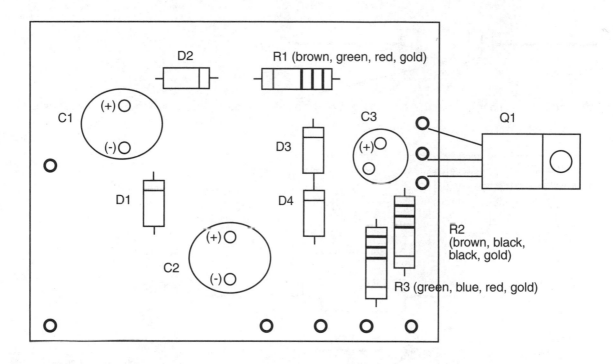

FIGURE 22-2
Component placement

5. Solder the leads (cut off any existing plug from the wire) from the transformer to the circuit board for D1 and C1, being sure to pull the wires through the hole in the box. Connect a black and red wire through the other end of the box, and solder to the board at the points shown. Mount the transistor to an aluminum heatsink, which will be at +65 volts and cannot touch anything metal except the transistor.

6. Plug in the AC adapter for about five seconds. Check to see if the LED lights up. Unplug the adapter and check if any components are getting hot. If any components are getting hot and/or the LED does not light up, you have a problem on the board.

7. If no components overheat, plug the adapter back in and measure the voltages shown in Figure 2-8.

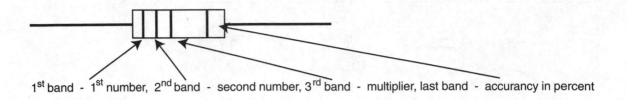

1st band - 1st number, 2nd band - second number, 3rd band - multiplier, last band - accurancy in percent

FIGURE 22-3
Resistor

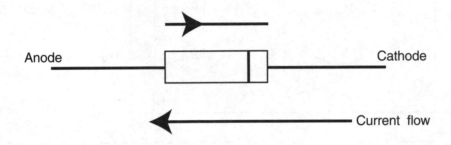

Anode   Cathode

Current flow

FIGURE 22-4
Diode

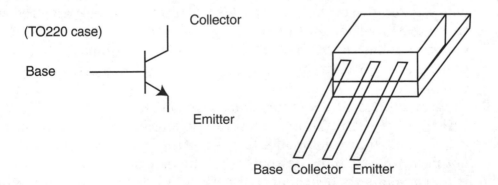

(TO220 case)   Collector

Base

Emitter

Base Collector Emitter

FIGURE 22-5
Transistor (TO220 case)

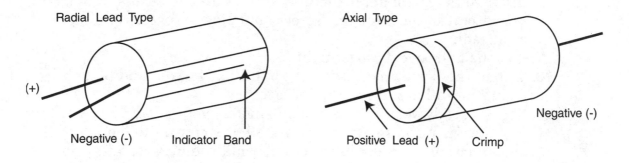

FIGURE 22-6

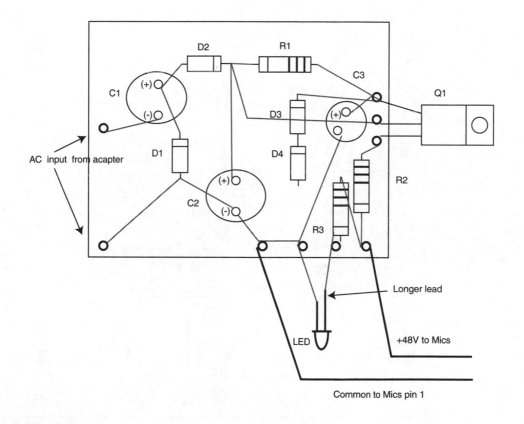

FIGURE 22-7
Underside wiring

8. If all voltages are approximately correct and all seems well, mount the circuit in the box using screws and plastic spacers or simply silicone adhesive.

9. Let the adhesive dry overnight.

10. Retest the box by measuring the output at the red and black wires, being careful not to short them together.

11. If all is ready, you are now ready to modify your patch bay by changing the wiring as shown in Figure 2-9. If you want to mount all this on a microphone plate in the studio wall, you may certainly do this, as shown in Figure 2-10. For each channel, you will need two 100-microfarad, 63-volt capacitors and two 5600-ohm resistors. Be extra careful of the positioning and polarity of the capacitors.

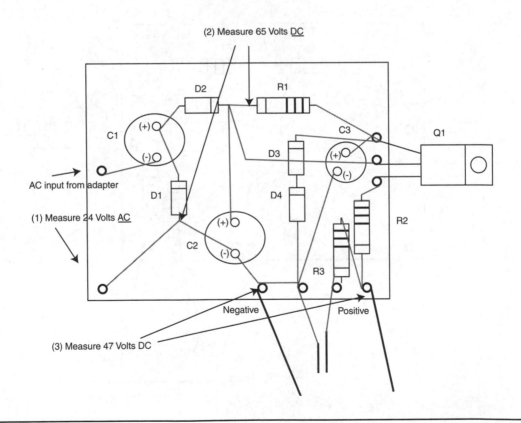

FIGURE 22-8
Testing and measuring

12. Be sure to test pins 2 and 3 of each XLR connector for 48 volts. You do not want to damage any dynamic or ribbon microphones by putting current through the internal transformers.

Component Identification

Resistors

1 = brown, 2 = red, 3 = orange, 4 = yellow, 5 = green, 6 = blue, 7 = violet, 8 = gray, 9 = white, gold = 5%

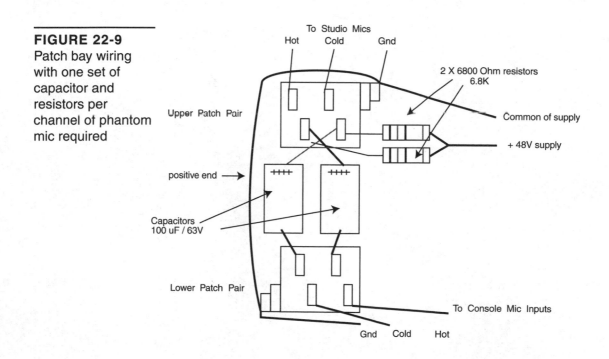

FIGURE 22-9
Patch bay wiring with one set of capacitor and resistors per channel of phantom mic required

FIGURE 22-10
Studio microphone
panel with one set
of capacitors and
resistorsper
condenser
microphone)

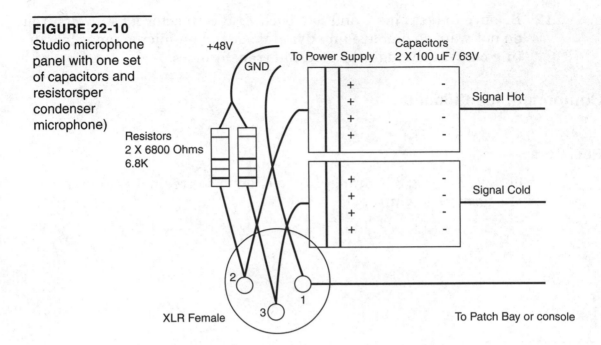

MAINTENANCE

CHAPTER TWENTY-THREE

Cleaning Chemicals and Applications

Proper cleaning chemicals are extremely important for keeping studio equipment in top condition. Misuse of the wrong cleaner can be disastrous. Too much solvent or cleaner can also be destructive rather than beneficial. Table 23-1 is a list of chemicals, starting from the most often used to the least often, and the components they are used to clean.

TABLE 23-1 Cleaning chemicals

Chemical	Areas of use
Isopropyl 99 percent pure alcohol	Fixed heads, rotary heads, rollers, dry faders, tape path components, circuit board edge connectors, and lenses in lights.
Lacquer thinner	Stubborn oxides on fixed heads, rollers, capstans, tape lifters, and other *metal* tape path components of a multitrack recorder. *Do not use on rotary heads of digital or video recorders!* The chemical will destroy the rotary head as well as the plastics.
Glass and plastic cleaner	Surfaces where a delicate but residue-free cleaner is required, such as panels, meters, screens, silk-screened console surfaces, and mirrors and light filters in lighting equipment.

(continued)

TABLE 23-1   Cleaning chemicals (continued)

Chemical	Areas of use
Soap and water	Natural rubber pinch rollers on multitrack analog tape recorders. Do not use on digital recorder rollers, as the water may migrate into the mechanism.
Toilet bowl cleaner	Ceramic capstans of multitrack recorders (usually white in color).
Contact cleaner	Patch bay jacks, cables, circuit card connectors, and switches.
Contact lubricant	Patch bay jacks, switches, wet-type faders, and potentiometers.
Wooden-handle Cotton swabs	Multitrack heads, tape paths, dry type faders, and digital recorder pinch rollers. Do not use cotton swabs to clean rotary heads, as the fibers will tangle up in the tiny heads and break them.
Nylon-handle doeskin swabs	Rotary heads on video recorders and digital audio recorders.
Compressed air spray can	For blowing dust out of components and mechanisms.

CHAPTER TWENTY-FOUR

Fixed Head Tape Recorders—Analog

The fixed-head multitrack recorder requires considerably more cleaning and attention than the digital recorder (see Figure 24-1). It has numerous exposed components in the tape path that can become contaminated by dust, oils, and tape particles. The tape path should be cleaned every time the tape is unloaded. The pinch roller and capstan require cleaning only when visible contamination is present. Some of these large machines can last for 10 years without requiring major maintenance if they are well cared for. Dirt is the biggest enemy of all equipment and must be kept out.

Figure 24-1 shows where most of the oxides and dirt build up in the tape path. Give these items the most attention for optimum performance of the machine. To clean a fixed-head multitrack recorder, follow these steps:

1. Clean the heads with an alcohol-dampened, wooden-handled cotton swab in a horizontal motion while working your way downward on the head. Use a new swab for each head (perhaps as many as three).
2. Clean the rollers and guides with another cotton swab working your way downward while rotating the guide if it is a rotary type.
3. Clean the capstan with a soft, lint-free cloth dampened with alcohol. Some machines can be tricked into spinning the capstan by inserting a business card into the sensor light where the tape normally travels. If you cannot do so, you may have to rotate the capstan by hand. If the capstan is ceramic and has stubborn residue, it can be cleaned with a ceramic cleaner such as a toilet bowl cleaner. If the capstan is metal and has stubborn residue, it can be cleaned

FIGURE 24-1
The fixed-head
multitrack recorder

ERASE HEAD   RECORD HEAD REPRO HEAD   CAPSTAN   PINCH ROLLER

with lacquer thinner. In either case, always reclean the capstan
with alcohol afterward to remove residue from the more aggressive
cleaners.

4. Clean the rubber pinch roller(s) with a wet, soapy cloth (20% soap)
   by rotating the rollers. If they do not seem to get any cleaner, you
   could use a rubber conditioner. *Never use alcohol on open reel pinch
   rollers!*

5. Demagnetize the heads and rollers by turning off the tape recorder
   and turning on the demagnetizer about three feet away. While making circular motions with the demagnetizer, come as close to each object in the tape path as possible without actually touching the objects. While still making circular motions, slowly move to other parts
   until finished. When finished, pull the demagnetizer back about
   three feet and switch it off.

This procedure should be done between each roll of tape and at the end
of each day of use.

CHAPTER TWENTY-FIVE

Analog Tape Recorder Calibration

Electronic Calibration (Routine Maintenance, Monthly If Desired or Required)

Calibrating a tape recorder in order to meet current industry standards is essential in order for the recorder to produce tape that will be compatible with other machines. Recorders that are not compatible create havoc for the studios receiving the resultant, nonstandard format at their location for re-mixing.

The first area of concern is the reproduction circuits, or repro and sync. The calibration of repro begins with an industry-standard test tape, which MRL laboratories produce. The appropriate tape for the required flux levels (magnetic strength) is not essential, but it is a good idea. The best way to choose an appropriate test tape is to consider your most common level settings. Most engineers reference to 250 and will specify levels in dB above or below the 250 reference.

Let's say you will be using 250 nano-webers per meter of tape length, or nWb/m. If you set your machine to this level, your *volume unit* (VU) meters will read 0 VU, and the output will be at the specification levels for the machine. If you need to calibrate to a different level, such as 320, you would simply calibrate the reproduce and sync levels for -2 at 250 and record at 0. The result would be 0 at 320 nWb/m. The levels are 160, 200, 250, 320, 420, and 630, and each is 2dB different in level (see Figure 25-1).

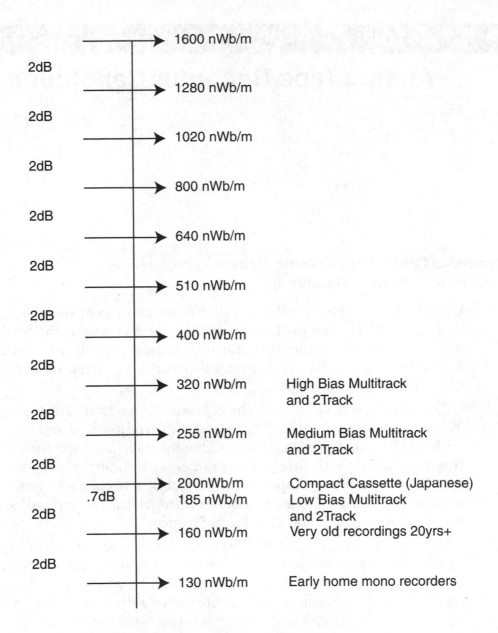

FIGURE 25-1
The various levels in dB corresponding to the flux levels on tape

The procedure for tape recorder calibration is as follows:

1. Clean the heads and check the path for any debris that would cause problems.
2. Load the test tape and rewind it to the beginning.
3. Set the machine to REPRO or TAPE mode and begin playing the tape.
4. The first tone is the 0VU or 0dB reference level, which is used to adjust the machine's repro level to provide a meter reading of 0dB. Adjust the repro level on each of the channels to maintain a reading of 0VU on the machine's meters.
5. The next tones are for checking and (if needed) adjusting the azimuth of the repro head.
6. The next tones start at the low frequencies and proceed to the highest. If the machine has low-frequency equalization adjustments, it may be adjusted in the area of 100 Hz on each channel. The high-frequency tones are more critical and may require a couple of passes per channel to get satisfactory results.
7. Set the machine to SYNC mode where the record head is switched to play mode. Steps 4 to 7 are repeated while adjusting the sync levels and equalization.
8. Unload the tape and clean the heads again. Then load a fresh tape.
9. Set the recorder to INPUT mode and feed a 1 kHZ tone at the required level to provide 0VU on the recorder's meters. You may need to feed only one channel at a time if the console cannot provide slating to all channels.
10. Set the recorder to REPRO and start recording while observing the readings on the VU meters. Change the frequency to 10 kHz and observe again. If the machine is only being touched up, you do not need to set the 3 dB over bias in the next step. You only need to slightly adjust the bias to make the meter read the same as the 1 kHz tone did.
11. To set the 3dB over bias point, you need to record a 10 kHz tone and adjust the bias either direction until a maximum level is achieved. You may need to reduce the oscillator level to be able to read this level. Then adjust the bias clockwise until the level drops by 3 dB. This is the 3 dB over bias setting (see Figure 25-2).

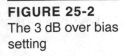

FIGURE 25-2
The 3 dB over bias setting

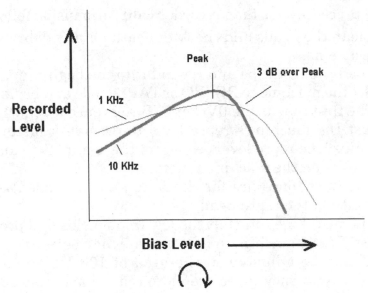

12. Set the generator between 1 kHz and 10 kHz, and adjust the high frequency record equalization so that the two frequency levels match.

13. Set the recorder to INPUT mode and the oscillator to 1 kHz. Adjust the oscillator level so the recorder's meters read 0VU. Then set the recorder to REPRO and adjust the record level for the same reading.

14. At this point, you have adjusted the equalization for the playback levels, sync levels, record bias, and record levels.

If the record equalization cannot be adjusted adequately, you can slightly adjust the bias to make up for the lack of adjustment range. Bear in mind that the bias works in reverse to the equalization adjustment. Less bias gives a higher frequency level. More bias reduces distortion.

If the response cannot be achieved or the heads have been lapped or replaced, a mechanical alignment is required, as described in the next section.

Head Alignment

The following adjustments are usually performed by a service person with a considerable amount of experience in performing these settings.

These adjustments are not usually required unless the heads have been replaced or lapped. *Please read the tape recorder's operation/service manual before performing these adjustments.* Some tape recorders have five adjustments per head, all of which can get you into a huge amount of trouble if done wrong. The first adjustment is to be performed on the playback head, followed by the record head (see Figure 25-3).

The adjustments are as follows and in order:

Height is the vertical adjustment of the head so that it is centered on the tape and is adjusted by all three screws equally (A, B, and C). At optimum, the levels are balanced on the first and last tracks.

Wrap is the adjustment of the head by loosening (D) and adjusting its angle. This is done so that the gap of the head is at its best contact point and is adjusted in the high-frequency range of the test tape for maximum output. At optimum, the high frequencies are stable and peaked.

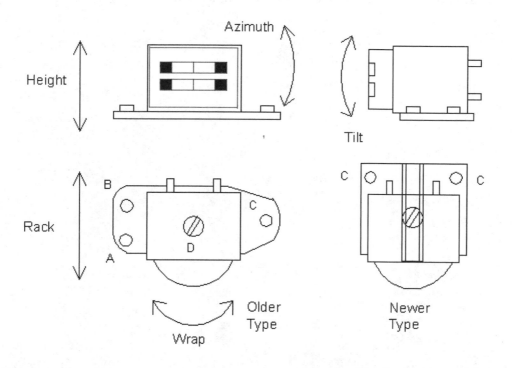

FIGURE 25-3
Older style head adjustments—Newer types only have Azimuth

Rack is the adjustment of the head for proper pressure against the tape (not often found on newer machines) using adjustment screw D. At optimum, the high frequencies are the most stable.

Zenith or *tilt* is the angle of the head relative to the tape contact. Misadjusting this can cause uneven and premature wear of the heads. This adjustment is usually done by a gage temporarily mounted on the transport; however, it can be performed by eye using a small flashlight and observing how parallel the head is with the guides.

Azimuth is the angle of the head in the same plane of the tape where the gaps are at 90 degrees to the longitudinal of the tape. This is a critical adjustment for keeping the machine compatible with others in the industry. When this adjustment is done properly, the phase relationship from channel to channel is correct and prevents cancellations at higher frequencies. The adjustment for this is screw C. At optimum, the best high-frequency response is achieved and nearby channels are in phase throughout *all* the frequencies on the test tape.

CHAPTER TWENTY-SIX

Rotary Head Tape Recorders—Digital

Properly cleaning the digital recorder can save costs in the long run. The caution to be taken with digital recorders is that the heads are extremely delicate. Care must be taken not to apply too much pressure with the cleaning swab.

Figure 26-1 shows the correct position for holding the swab while applying slight pressure and rotating the head counterclockwise. Note the angle of the swab. It is not necessary to remove the mechanism as shown.

The steps for cleaning are as follows:

1. If possible, load a known tape and get an error rate reading. You may have to get the procedure from a service center.
2. Clean the head with an alcohol-dampened doeskin swab.
3. Clean the track at the bottom of the head where the tape follows around the head with an alcohol-dampened cotton swab.
4. Clean the guides, tension arm, pinch roller, capstan, impedance roller, and loading arm with an alcohol-dampened cotton swab.
5. Allow the mechanism to dry and reinstall the cover.
6. Recheck the error rate. A lower value should be displayed.

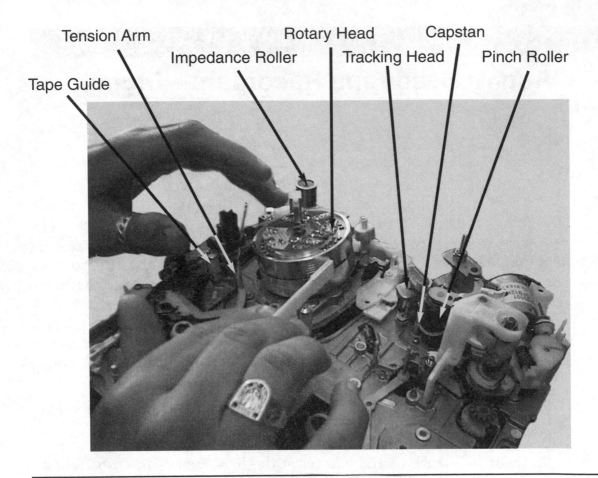

FIGURE 26-1
Cleaning a rotary head tape recorder

CHAPTER TWENTY-SEVEN

Amplifier Cleaning

Amplifiers are a hard-working part of the sound system and they generate heat when operating. They create air movement up and through themselves and thus pull dust and fibers along with the air. These particles form a blanket over the components that need to be cooled, and as a result, the amplifier will begin to run hotter over time. Eventually, it will fail if not maintained. Fan-cooled amplifiers are more prone to this thermal failure than convection-cooled ones because they maintain high flow of air (see Figures 27-1 through 27-6).

Convection-Cooled Amplifiers

Convection-cooled amplifiers require less maintenance than forced-air types because the heatsinks are on the outside and no fans are used. The cleaning procedure may require opening the top cover to clean some of the smaller heatsinks. Be careful to read the procedures listed in this chapter before attempting to clean the unit.

Caution: Wear a mask when cleaning equipment. The particles of dust and fibers that you inhale may contain dangerous bacteria.

Remove the amplifier from the rack or case and leave it *unplugged from the AC power for at least 2 hours*. This allows the power supply to discharge. A tremendous amount of stored power is contained in the filter capacitors, which are large metal cans usually located near the center of the chassis. They may have a plastic cover over them, and if this is the case, you probably don't need to worry about shorting them. They usually have two large screws and three or four large wires on each of

FIGURE 27-1
A suitable air
compressor for a
service shop

FIGURE 27-2
A typical night club
amplifier

FIGURE 27-3
Evidence of a lack of cleaning. Accumulated fibers in fan.

FIGURE 27-4
A blanket of fibers all over everything. Note the exposed terminals of the capacitors— DANGER— Capacitor store a HUGE charge. Avoid shorting them with any metal object!!

FIGURE 27-5
Spooge
everywhere!!

FIGURE 27-6
And more!!

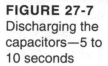

FIGURE 27-7
Discharging the capacitors—5 to 10 seconds

them. They must be allowed to discharge before any metal objects are brought near them in case of accidental shorts brought about by touching them. Such an accident would result in a rather frightening experience accompanied by sparks of molten metal shooting from the terminals.

If you are in a hurry or not sure if the power has discharged, you may want to discharge the amplifiers with a household incandescent lamp. To do this, you need to verify the bulb works and is on; then unplug it from the wall and attach a jumper clip lead to each plug pin. Clip one of the other ends of the clip lead to one capacitor screw and then hold the other clip lead end to the other capacitor screw. The bulb will light for a moment and then dim out slowly if a charge exists. Repeat the procedure for the other capacitor (see Figure 27-7).

To further clean a convection-cooled amplifier, follow these steps:

1. Clean the outer heatsink fins with a paintbrush and a vacuum cleaner.

2. Unscrew the top cover screws, being careful not to accidentally drop any screws into the chassis.

3. Use a can of Airjet, Dust Off, or any brand-name dusting spray and clean out the chassis around all the devices, being careful not to touch any of them.

4. Turn the amplifier over and pat the bottom plate so that any loose objects fall out.

5. Turn the amplifier back over and replace the top cover with the original screws.

6. Reconnect the amplifier in the rack or case and test it for a while (see Figure 27-8).

Forced-Air Amplifiers

The same rules of caution apply to a forced-air-cooled amplifier as a convection-cooled one regarding the stored charge in the power supply. Remove the amp from the rack and power source for at least two hours prior to servicing. You may discharge the capacitors in the same way as previously described.

This type of amplifier can be a little more stubborn to clean. The dust needs to be blown out the way it came in. You may have to observe the direction of airflow prior to disconnecting it from the power and system,

FIGURE 27-8
Convection-cooled
amplifier

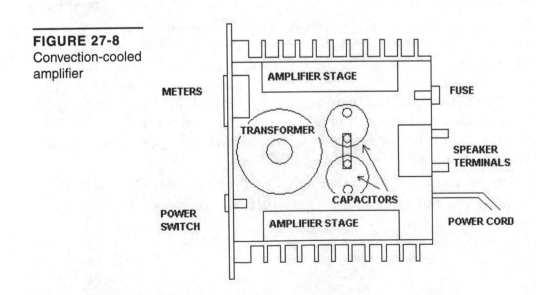

but do not remove the covers to do this. This type of amplifier requires the covers for cooling. A can of compressed air may have enough force to clean the heat tunnel; if not, you may need to use an air compressor with 40 to 50 pounds of pressure. Once you know the direction of airflow and have left the amplifier disconnected for at least two hours or discharged the capacitors, you can clean out the heat tunnel opposite the original airflow. The exit end of the heat tunnel may not be easily accessible so you may need to take the top cover off the amplifier to get to it.

As with the convection-cooled amplifier, you need to clean the dust from the internal components as well. If the cover is still on, you need to remove it for access to the circuits inside the unit. Take care not to touch anything inside the amplifier. Blow out all accumulated dust and brush off the fan blades. Then follow these final steps:

1. Turn the amplifier over and pat the bottom plate to knock out any loose objects.
2. Reinstall the top cover and power the unit up.
3. Apply a signal and observe that the fan(s) come on when needed as before (see Figure 27-9).

You may need to set a schedule for cleaning for your equipment, and the amplifiers are one of the most important to consider. Heat will destroy an amplifier just as sure as overheating will destroy your car engine.

FIGURE 27-9
Forced air type amplifier

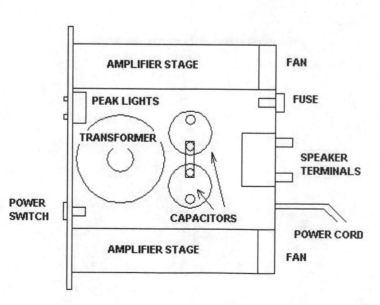

CHAPTER TWENTY-EIGHT

Computer and Processor Cleaning

Computers, certain amplifiers, and some audio processors use switching power supplies with fans (see Figure 28-1). The most common fault with these items is overheating. This occurs because the fan has become clogged with debris or the fan has seized up and is no longer delivering cool air to the unit. Cleaning out the fan and power supply using compressed air easily prevents this situation. *Be sure to unplug the equipment before opening it.*

The first step is to remove the fan and lay it face down. Fans usually have a small circular label on them at the stationary frame end. This label can be carefully peeled back, the gasket pellet lifted, and a drop or two of machine oil can be added to the bearing (see Figure 28-2). Then without getting any oil on the adhesive, the label can be reinstalled along with the pellet. The fans should be lubed every year, and the power supplies should be cleaned every 1,000 hours. Sometimes the fan gets noisy as the bearing gets dry.

If convenient, lay the fan face down for a while to let the oil migrate down into the bearing before reinstalling it. Avoid getting any oil on the adhesive.

All removable circuit cards should have the edge connectors cleaned with a high-quality contact cleaner/restorer while the unit is open. Be sure to put items back where they came from—*exactly!!*

FIGURE 28-1
Cooling fan

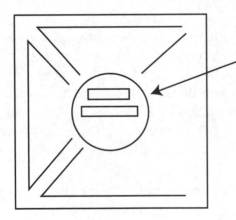

Peel back this label and add oil.

FIGURE 28-2
Adding some machine oil to the fan

CHAPTER TWENTY-NINE

Mixing Console, Pot and Fader Cleaning

The procedures for removing the channels and faders may vary slightly depending on the console and manufacturer. Whenever possible, refer to the manufacturer's service manual, which should always be obtained with the console. Most large consoles are built in the same general way (see Figure 29-1).

FIGURE 29-1
A typical large console

To clean the console, follow these steps:

1. Lay a towel on the console surface to avoid scratches.
2. Remove the screws, holding down the fader strip (see Figure 29-2). Usually one screw is at the top and another is at the bottom.
3. Extract the module from the body of the console and lay it on the towel. Sometimes an extractor tool is required (see Figure 29-3) and occasionally screws must be removed at the rear of the console (see Figure 29-4).
4. Remove the fader knob (see Figure 29-5) and then the screws (see Figure 29-6), holding the fader in place through the front panel.
5. Lay the fader on its side so that you can access the tabs or screws holding it together (see Figure 29-7).

Figures 29-8 through 29-12 illustrate the method of disassembling and cleaning a fader unit.

FIGURE 29-2
Removing the screws that hold in the modules

FIGURE 29-3
Using an extractor
tool to lift the
module

FIGURE 29-4
Fader and
connections

FIGURE 29-5
The fader knob
removed

FIGURE 29-6
Removing the
fader screws

FIGURE 29-7
The detached
fader

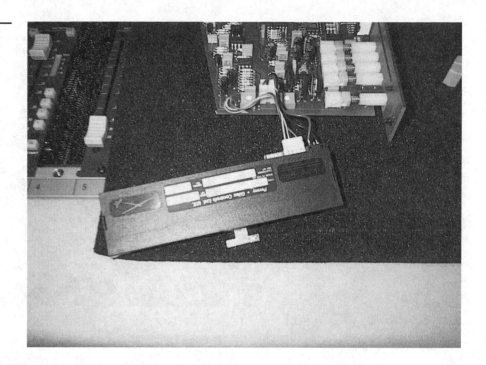

FIGURE 29-8
The fader unit

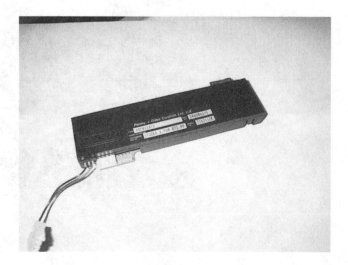

FIGURE 29-9
Opening a plastic
body type fader

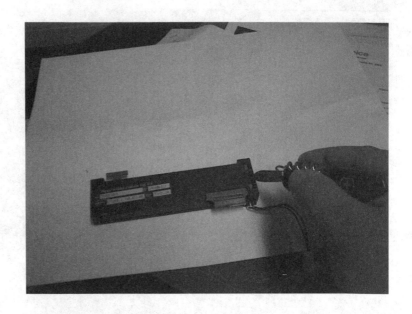

FIGURE 29-10
Lifting off the
cover

FIGURE 29-11
Cleaning the conductive print with pure 99% alcohol soaked cotton swab

FIGURE 29-12
Applying oil to rails after cleaning them

These figures show the sequence of disassembly, cleaning, and lubrication of one of the less expensive P&G faders. The top is released by lightly pressing inwards on the tabs at the ends. Turn the fader over and lift the cover off. Be sure to clean the circuit board conductive tracks (in black) with alcohol only (no contact cleaner). Add a drop of oil on the chrome slide rails and reassemble the unit. Do not let any oil get onto the circuit board. Make sure the slider is in the center of the fader when you reassemble the unit.

Figure 29-13 is another model of a P&G fader. Much more care must be taken when servicing this fader. Only remove the screws from the end with the wires. A square nut will fall out when you remove the small metal plate that holds in the dust cover. Always work over a clean, light-colored surface when servicing any fader. Always remember to reinstall everything the way it came apart. Make drawings or label the parts; do whatever you need (see Figures 29-14 through 29-17).

Note that the spring wipers in the body face inward. This allows you to slide the circuit boards back into the shell without damaging them. If the wipers were facing outward, you would not be able to reinstall the

FIGURE 29-13
Metal body type
P&G fader

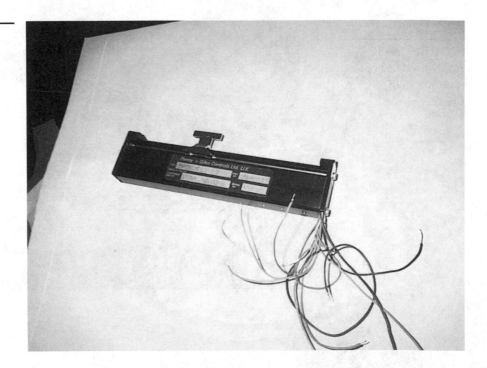

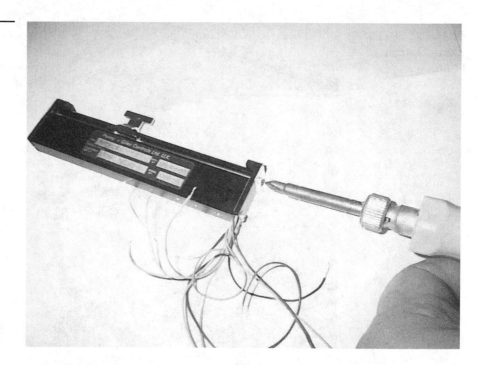

FIGURE 29-14
Opening the proper end

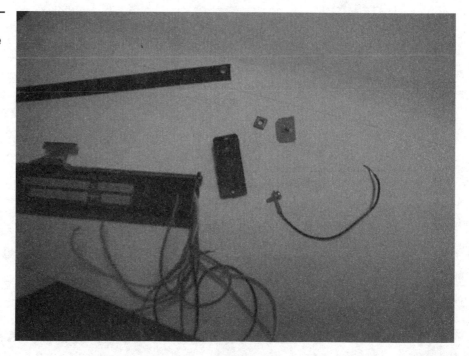

FIGURE 29-15
Careful not to lose any small parts!

FIGURE 29-16
Extract circuit
board

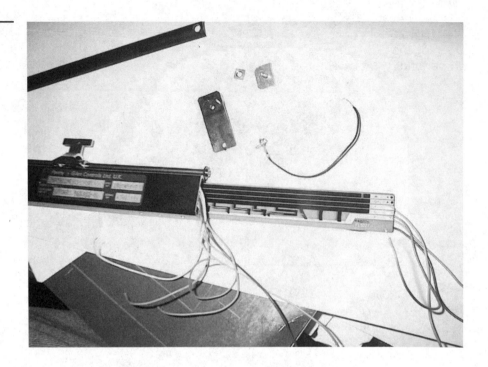

FIGURE 29-17
Clean conductive
print and rails, oil
and reassemble

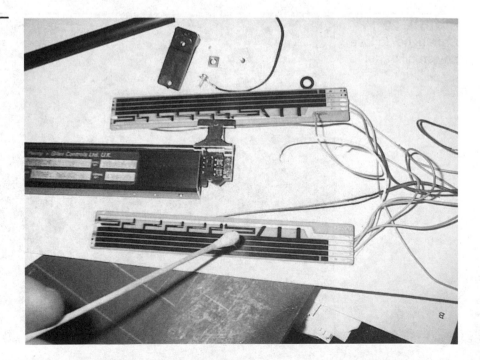

boards. As before, clean the conductive tracks with alcohol and lube the chrome rail.

Figure 29-18 is a cross-fader. This cannot be opened because folded metal tabs hold the circuit board onto the metal body. The procedure for cleaning a cross-fader is to spray a high-quality control cleaner inside it while vigorously operating the handle. Then inject some fader lubricant afterward, and again operate the slide arm unit a few times to work in the lubricant. The unit will feel a little different for a while as the lubricant settles in.

Rotary pots are cleaned in a similar method as cross-faders in other circuits. Rotary pots usually have their openings at the bottom near the circuit board. You may have to make a bend in the spray straw of the cleaner to be able to aim into the opening of the control. As with linear action controls, the rotary controls require lubricant after cleaning. Be sure not to miss this step, as they will self-destruct without lubrication.

Note the opening for the arm. Spray the cleaner and lubricant on the sides and not the center. Bending the straw of the cleaner with a little heat will help (see Figure 29-19).

FIGURE 29-18
A cross-fader for a
DJ mixer

FIGURE 29-19
Insert straw with
bend towards
sides and apply
one or two short
squirts of cleaner

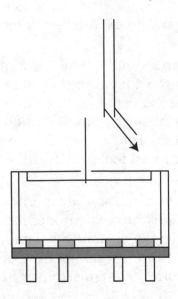

When servicing a console or multitrack recorder, it always reduces costs to use extender cards so that the machinery can be operated outside the console or tape recorder. This also allows for test equipment such as scopes to be attached while the unit is operating (see Figure 29-20).

FIGURE 29-20
Extender cards
in use

Extender Cards

CHAPTER THIRTY

Lighting Equipment Cleaning

Lighting apparatus live a rather hard life in regards to a lack of proper maintenance and duty cycles. They live in dusty environments where carpet fibers are their biggest enemy. These carpet fibers stick to everything hot in the units, including the vital fan. This results in a blanket covering all of the objects, which need to get fresh, cool air.

The most important piece of lighting maintenance gear is the air compressor. If you are a club owner or DJ who is responsible for keeping the equipment in top running order, you should have an air compressor and blow gun fitting. The equipment needs regular inspection.

The most sensitive parts of lighting fixtures are the voltage and the bulb. If the fixture uses a discharge-type bulb, as many as 8,000 volts may be used to start it. *Be sure to disconnect the mains or unplug the fixture prior to opening it for cleaning.* The bulb cannot be touched with your fingers. The oils from your skin will stick to the surface of the bulb and, when heated, will prevent it from heating evenly, resulting in an explosion. The gas-discharge bulbs cost anywhere from $300 to $400 and make a nasty mess in the fixture if they detonate. This also applies to halogen bulbs.

The bulb should only be cleaned with air and, if necessary, a paintbrush. Use air around the bulb and on the fan, which will be in the immediate area. Past the bulb is the first lens, which should be cleaned with 99.9 percent pure isopropyl alcohol (not rubbing alcohol). All the lenses in the unit can be cleaned with alcohol, but do not clean mirrors or color wheels this way. Color wheels and mirrors should be cleaned with a glass cleaner. The color filters are subjected to a lot of heat and will be unable to withstand a harsh cleaning chemical.

Use the paintbrush and air to clean the shutters and other mechanical devices in the unit (see Figure 30-1). The chrome-plated slides may require a wipe with alcohol to remove baked-on lubricants. Use a pencil or dry graphite to lubricate these, as any oil will attract more dust. Rub the pencil along the rails and on surfaces. The pencil is actually made of graphite and not lead, and the graphite is very slippery.

Figures 30-2 through 30-4 show some of the components in a typical lighting fixture. This particular unit gives basic colors and patterns in sweeping positions in the room.

Figures 30-5 and 30-6 show some accumulations of dirt and fibers in the fan and heat areas.

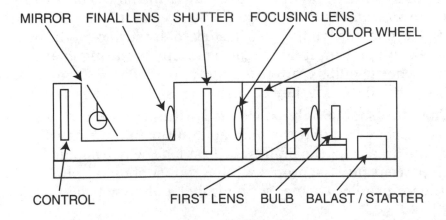

FIGURE 30-1
Basic internal components of a lighting device

FIGURE 30-2
The lamp area
and control circuit

FIGURE 30-3
The pan and tilt
mirror with final
lens

FIGURE 30-4
AVOID touching
the bulb with bare
fingers!

FIGURE 30-5
Dust Bunnys
(Spooge) in the
fan area

FIGURE 30-6
This stuff needs to be brushed out of the fan

SERVICE AND REPAIRS

CHAPTER THIRTY-ONE

Speaker Damage and Repairs

This section covers diagnosing and repairing speaker systems that are no longer in warranty. Any speakers under the manufacturer's warranty should only be serviced by factory-authorized personnel.

A speaker is a form of a linear induction motor that has low mass and high torque. Sound is created by the voltage changes that are applied to it. The cone size and weight determine how the speaker will respond and the frequencies it will respond to.

Many things can cause speaker damage. A few of these are overdriving, amplifier failure, system faults, system misuse, and power problems. Most manufacturers provide an extensive warranty for their speakers, but in most cases they will not cover burned voice coils, which are almost always a result of overdriving or faults in the amplifier system. If your speakers are in warranty, you cannot open them for inspection, as this is considered unauthorized service and will void most warranties.

The speaker driver consists of several components: the frame, magnet, armature, voice coil, spider, cone, dust cap, surround, gasket, and tinsels (see Figure 31-1).

Testing

If no sounds come from the speaker, the first test would be to use a 9-volt battery to click-test the speaker. Apply the positive of the battery to the speaker cabinet's positive terminal and the battery's negative to the cabinet's negative terminal while observing the speakers. If the woofer is in

FIGURE 31-1
Speaker
components

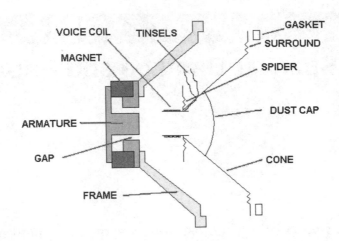

VOICE COIL · TINSELS · GASKET · SURROUND · MAGNET · SPIDER · DUST CAP · ARMATURE · GAP · CONE · FRAME

question, it can be observed visually, whereas if the midrange or tweeter is in question, it has to be listened to. The application of the battery to the terminals should be momentary so that you do not damage the speakers or battery because about 1 ampere of current will flow when connected to an 8-ohm speaker, which has about 6 ohms of DC resistance in the coil winding. The woofer should move outward, and a thump should be heard as the battery is applied. When the battery is released, another thump or crack will be emitted, and the speaker will move back inward.

In the event that the speaker does respond, the fault will most likely be in the wiring to the amplifier or in the amplifier itself. If the speaker does not respond, the woofer or suspect speaker should be removed from the cabinet, observing the wire colors to each terminal, and the battery should then be applied to the speaker directly in the same fashion. Beware that older JBL woofers respond backwards to most other speakers; they are actually wired in reverse phase to the rest of the cabinet.

Now apply the battery to the terminals and observe the reaction. If a response occurs and it sounds normal, the fault is in the crossover in the cabinet or it may be a blown fuse. If no response occurs or if you hear a scraping or distorted sound, you need to continue looking at the speaker for faults.

At this point you should use an ohmmeter set to read 200 ohms or less and, if possible, it should ring if continuity exists in the circuit. Apply the voltmeter to the tinsel leads where they meet the cone at the rivets. If a reading occurs, the tinsels need to be replaced. If no reading takes place, the problem is farther in the speaker and the next step is to remove the dust cap.

The dust cap can be removed by applying lacquer thinner to the adhesive (in order to soften it) around the dust cap where it meets the cone. This will take three or four applications without letting the adhesive re-dry. When the adhesive is soft enough, carefully score the edge of the cone so that its paper will not tear when removing the cap. Slightly cut the adhesive so that you can insert the edge of a knife under the cap. Then begin lifting the cap off very gently while it is still moist with lacquer thinner. The cap should lift off slowly as the glue releases. Be careful not to drop any particles into the speaker while the dust cap is off.

To test the voice coil, gently scrape the varnish off the voice coil wires, which are glued to the side of the cone, where they come out of the center to meet the tinsels entering the cone at the rivets. Measure these wires again with the ohmmeter. If you get a reading or tone on the meter, you need to resolder the connections at the rivets because the fault is most likely a poor connection where the tinsels meet the voice coil wires. If no reading or tone occurs, the fault is in the voice coil, which must be cut out. Do this only if the speaker is *not* under warranty.

Cutting out the voice coil and cone is the final phase of the speaker's inspection and renders the speaker inoperative. If you are unsure that you have completely tested your speakers properly or are not certain that your speakers are out of warranty, you should *not* proceed any further. Instead, send your speaker to an authorized speaker re-cone or repair center for further service.

You may think you would be better off just replacing the speaker unit. Bear in mind that you save about 70 percent by rebuilding the speaker. The worn and damaged parts all get replaced in the re-coning procedure.

Cutting out the Cone

To continue with this non-warranty speaker service, you need to cut out the cone and voice coil assembly for inspection. Once you cut the cone, the speaker is rendered inoperative.

Cutting out the cone is the point of no return for the re-cone commitment. It pays to double-check your tests before this step. Once you are certain that no continuity exists in the voice coil, you may proceed to cut the cone out as shown by the dashed lines in Figure 31-2. This exposes the spider, which is the yellowish round bellows like device at the bottom, so that it can also be cut out and release the voice coil from the gap. Unsolder the tinsel leads and save them for possible future repairs. Keep in mind that as long as you have spare tinsels, you will never need them, at least according to Murphy's law.

Cleaning

Carefully remove the voice coil and spider assembly out of the magnet and frame, being careful not to rub the edges of the voice coil so that debris does not fall into the speaker. Once the voice coil is free from the magnet, use masking tape to cover the gap where the voice coil was so that no dirt or particles fall in. Apply a lacquer thinner to the remaining glued-on paper in the frame and allow the glue to soften. Then, using a scraper, scrape off the old paper and glue from the frame where the spider and cone were fastened. When the frame is clean, remove the tape and carefully use a source of compressed air or duster spray to clean out the gap.

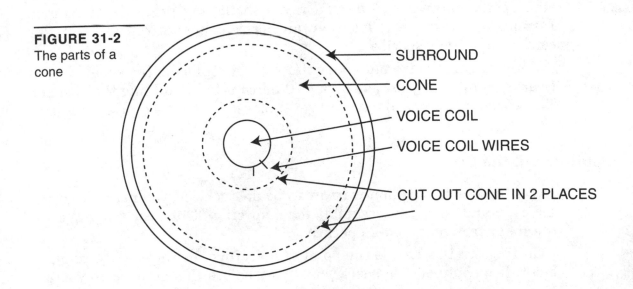

FIGURE 31-2
The parts of a cone

SURROUND

CONE

VOICE COIL

VOICE COIL WIRES

CUT OUT CONE IN 2 PLACES

You will likely get a set of spacer shims with your cone kit. Wrap some tape around one of the spacer shims with the sticky side facing out. Insert it into the gap while scrubbing up and down, working your way around the gap to clean off any stubborn dirt (see Figure 31-3). If no shims have been supplied, you may use some thin cardboard about as thick as a playing card or cut up a real card. Use some compressed air to blow out the gap again to be certain no dirt remains.

Assembling the Kit

You are now ready to install the new cone kit, which may not be assembled. If this is the case, follow the steps below; otherwise, skip this phase.

1. Slide the spider onto the coil from the top where the two wires extend. Keep the wires from bending and getting caught under the spider until it sits where the old one did on its coil (see Figure 31-4).
2. Carefully insert the cone onto the voice coil where it meets the spider so that the rivets are on the same side as the wires. Between $1/8$ and $1/4$ of an inch of the voice coil should be sticking up through the junction.

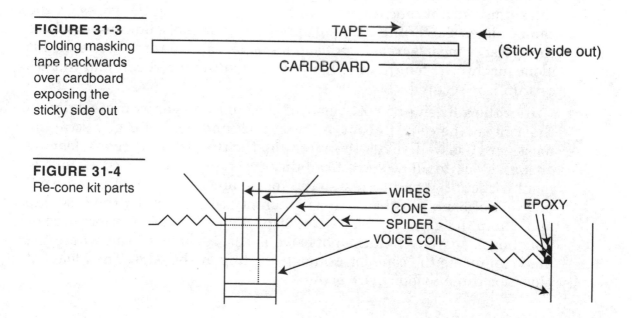

FIGURE 31-3
Folding masking tape backwards over cardboard exposing the sticky side out

TAPE
CARDBOARD
(Sticky side out)

FIGURE 31-4
Re-cone kit parts

WIRES
CONE
SPIDER
VOICE COIL
EPOXY

3. Lower this assembly into the gap until the spider just touches the frame. The cone's surround should also barely touch. If it does not, the spider may be upside down.
4. With the components sitting just right, insert the shims to keep the coil straight. Recheck the height and apply a bead of carpenter's glue or epoxy around the coil where the cone and the spider meet it, being extremely careful not to drip any into the gap.
5. Let this assembly sit overnight.
6. After the adhesive has set, remove the shims and pull the cone assembly out of the frame.

Installing the Kit

Before applying any adhesive, try a dry fitting a couple of times to make sure the replacement kit fits properly. The cone kit should slightly touch the spider and the surround at the same time when lowered in.

When you are happy with the fit, apply the adhesive recommended by the manufacturer around the frame where the cone and the spider make contact. Apply the adhesive to the same mating places on the cone surround and spider. Be sure to spread the adhesive evenly.

Install the shims and lower the cone kit into the gap, guiding it over the shims until it touches down in the frame, keeping the wires on the same side as the terminals. Apply pressure to make solid contact where the surfaces meet, spreading the adhesive for best contact. Leave the kit alone for three to four hours when you are happy with the fit and you are certain it is secured.

After the adhesive has set, remove the shims and gently press down on the cone at the center about $1/4$ of an inch and listen for any scraping noises. If all is well, solder the wires and the tinsels to the rivets, leaving enough slack to allow some flexibility through the entire cone's movement without tightening. Install the gasket and dust cover with more adhesive. Set some weight (such as an empty coffee cup) on the dust cap and clamp the gasket in place (clothespins will do). Leave it for three or four hours again. When the adhesive is set, reapply a bead where the dust cap meets the cone for complete sealing at the edge. Then leave it alone for three to four more hours.

When the adhesive is set, install and test the speaker at a low to moderate volume. Remember to rotate the woofers in the cabinet every three to four years by 180 degrees to relieve stress.

If the speakers have rotted foam surrounds but are still functioning, you can replace the surrounds and dust caps only. Several companies supply the surrounds and the caps as individual items.

The first step is to remove the speaker unit from the cabinet and clean out any loose debris (see Figure 31-5).

The second step is to remove the old material from the cone and basket. The mounting gasket needs to be replaced first. Do this by prying up the gasket with a knife, working your way along the edge (see Figure 31-6).

The next step is to prepare to remove the dust cap by wetting the edge with lacquer thinner or acetone solvent to soften the adhesive (see Figure 31-7).

Insert a knife between the cap and the cone, being careful not to perforate the cone. You should barely feel the dust cap begin to lift. Keep the glue moist with the solvent (see Figure 31-8).

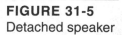

FIGURE 31-5
Detached speaker

FIGURE 31-6
Removing the
gasket

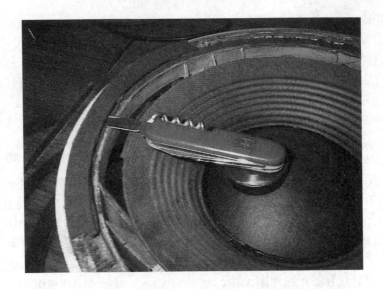

FIGURE 31-7
Softening the
adhesive to
remove the dust
cap

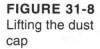

FIGURE 31-8
Lifting the dust
cap

Now clean the old foam off of the back of the cone. Usually, it just rubs off without solvents (see Figure 31-9).

Install paper shims into the voice coil gap to center the cone and keep it in place (see Figure 31-10). Pull the new foam edge onto the cone so that it can be secured from the back and center it.

Apply an adhesive (such as contact cement) between the foam surround and the cone. Spread it evenly with a cotton swab as you move around the cone. Then press them together, keeping the surround centered on the cone. You may have to turn the speaker over and view it from the top to ensure proper centering. The foam will curl and wrinkle slightly as the adhesive causes it to expand, which is normal. Press down on the foam surround and the cone every couple of minutes or so until they stay in place (see Figure 31-11).

Now apply an adhesive to the surround and frame. Spread with a cotton swab and press down with your fingers. As previously, the foam will lift as the adhesive causes it to expand. Press down on the foam to frame it all around while keeping the foam surround centered (see Figure 31-12).

FIGURE 31-9
Removing the
foam

FIGURE 31-10
Paper shims
holding the cone
in place

Press down every couple of minutes or so until it stays, as shown in
Figure 31-13.

Rub off any loose particles of the old surround so that it will lay flat on
the frame when mounted (see Figure 31-14).

FIGURE 31-11
Adding adhesive

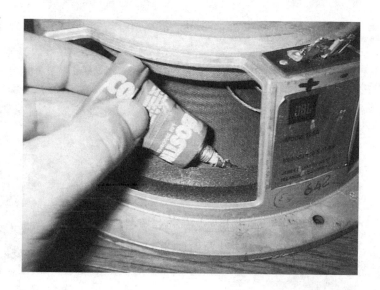

FIGURE 31-12
Adding adhesive
to frame or basket

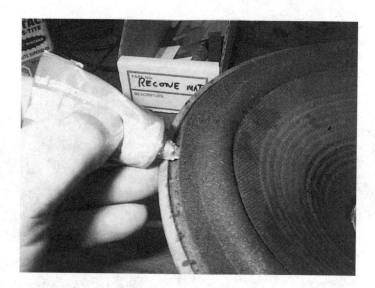

When the gaskets have been reinstalled and are fitting in a stable manner with the holes and the bolts aligned, set the speaker into the cavity upside down so that the weight of the speaker will help set the gaskets. Let the adhesive set overnight (see Figure 31-15).

FIGURE 31-13
Pressing the
surround and the
frame together

FIGURE 31-14
Removing the old
glue from the
gaskets, then
re-gluing them
down

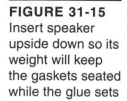

FIGURE 31-15
Insert speaker
upside down so its
weight will keep
the gaskets seated
while the glue sets

When the adhesive is set, remove the speaker from the cabinet and pull out the gap shims. Test the speaker with a 9-volt battery. Set the dust cap in place and apply a bead of adhesive along the edge where it meets the cone. Set a weight such as a coffee mug on it and let it set for at least three hours before reinstalling in cabinet. Apply another bead of adhesive between the dome and the cone and let it set again. If desired, you may want to spray a light coat of flat black paint on the face side of the cone to make it look new. Do not use the speaker at high power for at least six hours.

Repairing Horn Tweeters

To determine whether the horn driver has been blown, apply a 9-volt battery momentarily to the driver's terminals. You should hear a pop sound if the driver is working. If no sound is made, the driver is more than likely blown open.

In most cases, the driver can be removed from the front panel with the horn assembly. The driver most often unscrews from the horn on a threaded fitting. The horn is dismantled by unscrewing the three or four screws holding the diaphragm onto the magnet assembly. Locating pins hold the diaphragm in position and replacing the diaphragm is a simple procedure. Make certain that no debris is in the magnet gap. If it contains some, the magnet gap can be cleaned the same way as a woofer gap. Be sure to use removable loctite 242 to lock the threads when re-assembling the driver onto the horn (see Figures 31-16 through 31-20).

FIGURE 31-16
A two-way PA
speaker with horn

FIGURE 31-17
Horn assembly
extracted

FIGURE 31-18
Driver unscrewed
from horn

FIGURE 31-19
Diaphragm
removed from
magnet

FIGURE 31-20
Non-permanent
thread locking
compound

CHAPTER THIRTY-TWO

Amplifier Diagnostics

Amplifiers exhibit many different types of faults when they fail. Some of the most common are DC offsets, total shutdown, overheating, oscillations, and an unbalanced drive.

Most amplifiers have speaker protection circuits designed to disconnect the amplifier from the speakers during failure mode. This is sometimes indicated on the front panel by a *light-emitting diode* (LED) that says protect. This circuit may be activated by overdriving, overheating, a shorted speaker line, DC offset, or other speaker-damaging voltages.

If the amplifier does not have a protection circuit and a fault is present, a rather loud hum or buzz sound may occur, accompanied by a loud thump or bang when the amp is shut off. The easiest way to diagnose this type of failure is to get an incandescent desk lamp and turn it on and then unplug the cord while the lamp is still on. Using a pair of alligator test leads, attach the power plug of the desk lamp to the speaker terminals instead of the speaker. Turn on the amp with the volume down and see if the lamp lights up at all.

You can also use a voltmeter and measure for DC voltage on the speaker terminals. If it reads less than 1 volt, switch the meter to read AC Volts. More than likely a DC voltage will occur if the speaker made a loud noise when the amp was shut down. This information should be passed on to the technician who will repair the amplifier.

If the amplifier shows a heat fault or gets abnormally hot, more than likely a cooling system failure has occurred. The amplifier may be a forced-air-fan type, and if so, check to see if the fan is running when the amp gets hot. If it does not, the fan has seized up and needs to be replaced. If you replace the fan yourself, make sure that the fan is the same type and requires the same power source, which is printed on the label of the fan. Also make sure the fan is installed so the air will flow in the same direction as it would have previously. Arrows on the body of the fan indicate its rotation direction and airflow. If the amplifier does not have a fan or is showing heat problems, it probably has developed a fault in the bias circuit and needs to be repaired by a technician.

A humming sound or radio station appearing in the speakers could be the result of interference in the cables. Refer to Chapter 33, "RF Interference Cures" for the possible fix.

A buzzing sound can be caused by a failing power supply in the amp, lighting fixtures, or controllers on the same power circuit. See the section on technical power in "Basic Formulas for Calculations" for more information.

In many cases, the equipment is blamed for noises and interference conditions in the power lines that cause trouble. Always move the equipment to different power sources and test it before sending it in for servicing. Also document the application and situation for the technician so that the troubles can be repeated in the repair shop.

A blown fuse indicates that the amplifier has been driven beyond its maximum output, has a shorted speaker wire, or has an internal failure. To test for this, remove the speaker wires from the output jacks and replace the fuse with the *same type only*. Then disconnect the inputs to the amplifier and turn it on. If the fuse blows again, the amp has a severe problem. If the fuse does not blow, measure for a DC voltage at the output terminals and connect an incandescent lamp to them. If no DC voltage is present and the lamp does not light, switch off the amp and reconnect it to the speakers and the system. Very carefully turn up the volume with some audio program and listen for a distorted sound.

A distorted sound indicates that either the amp has a failure or the speaker is damaged. Reverse the speakers from channel to channel to see if the distortion moves to the other speaker. If the distortion moves, the amp has a fault inside. If the distortion stays, the speaker is damaged.

A buzzing noise can indicate that the amp is receiving interference from the cables going into it. This can be tested by simply unplugging the input connectors to the amp and listening to the speakers. If the buzzing continues, the amp may have an internal power supply fault and will require a technician to service it.

A hum can be the result of one of two problems: too many grounds in the system or an oscillation occurring in the amp. The previous procedure can determine whether the problem is internal or external to the amp.

Amplifiers have a great deal of current stored in their power supplies, so caution must be taken when dealing with them. Do not reconnect the amplifier to the system until some basic tests are done to make sure that no speaker damage will occur. This applies to amplifiers that are returning from being serviced as well.

Soldering, Cable Repair, and RF Interference Cures

Properly grounding the equipment in the studio is important and will provide the best *radio frequency* (RF) protection for the signals. If, however, the grounding is done properly and the RF interference (from lower AM bands) is still getting in, more involved procedures are required. Let's first look at proper soldering procedures.

A 35-watt-minimum soldering pencil is essential. Twenty-five watts is not enough for connectors. The connector will cool off the tip of a 25-watt pencil, and the connection will be poor because the solder did not properly melt and flow. A $^{60}/_{40}$ resin core solder is also required. Avoid an acid core solder! A damp sponge for tip cleaning and a proper soldering iron holder are also essential.

The soldering iron tip should be placed at an angle to the connection to best transfer as much heat as quickly as possible. In some cases, a fresh solder will be necessary to release the old connection and is always necessary to start a new connection.

A couple of seconds after introducing the iron to the connection, add some solder to coat the surface areas to be connected (see Figure 33-1).

As done previously, after holding the tip of the iron against the wire for a couple of seconds, add enough solder to run up the wire to the insulation (see Figure 33-2).

FIGURE 33-1
Adding solder to the terminal (Tinning)

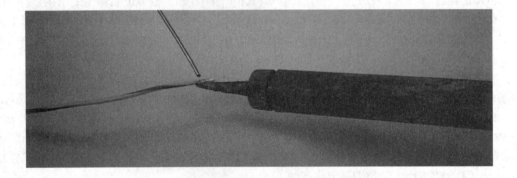

FIGURE 33-2
Tinning the wire—let the solder flow into the wire strands

Place the wire where it must be connected and use the iron tip to melt the solder and alow the wire to sink into the solder. Enough solder should coat the surfaces so that when the iron touches the two, the solder will melt and flow together. Set the soldering iron tip against the two and hold for two to three seconds or until the solder melts. Once melting occurs, remove the iron and continue to hold the two together for 10 to 20 seconds so the solder sets shiny. Resting your wrists on the work surface will prevent your hands from shaking and make for more clean-looking work (see Figure 33-3).

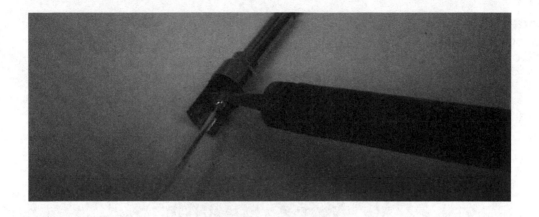

FIGURE 33-3
Re-melt the solder in the terminal and allow the wire to sink into it

A *ferrite bead* installed on the signal line(s) but not the ground drastically cuts down on RF interference in the system from cables and connectors. This procedure is most effective in the processor and monitoring loops (see Figures 33-4 and 33-5).

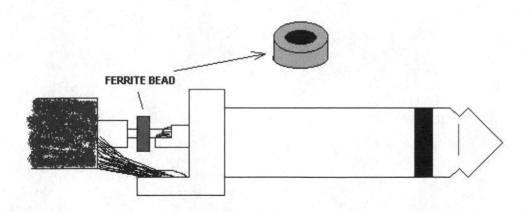

FIGURE 33-4
A ferrite bead installed

FIGURE 33-5
Many beads for
severe RF troubles

CHAPTER THIRTY-FOUR

Processors and Outboard Gear Service

Processors such as equalizers and compressors are usually very reliable. The most common failures are switch and control noise. In the most expensive processors, the controls are usually sealed and not designed to be cleaned. If you do not have a readily available source for the replacement controls, you can clean them if you are careful. The first thing to do is remove the controls or subassemblies from the unit (see Figures 34-1 and 34-2).

FIGURE 34-1
Processor
modules

FIGURE 34-2
Modules in
chassis

Next, drill two tiny holes where the leads extend from the body of the control, one above and one below. Be absolutely certain to drill only through the casing. Use a drill for ink cartridge refilling if possible so that you feel *precisely* when the drill tip goes through (see Figures 34-3 and 34-4).

FIGURE 34-3
Holes drilled for
adding cleaning
fluid and lubricant

FIGURE 34-4
Adding fluid

Squirt two or three full pumps of cleaner lubricant into the control, and rotate the control back and forth 10 to 20 times. Then align a compressed air straw into one hole. Protect your eyes and blow out the excess cleaner. Rotate the control a few more times and reinstall it in the unit.

APPENDIXES

Symbols and Keywords

Hundreds of different symbols are used for electronics and signals. The following is a list of some of the most common ones used in studios.

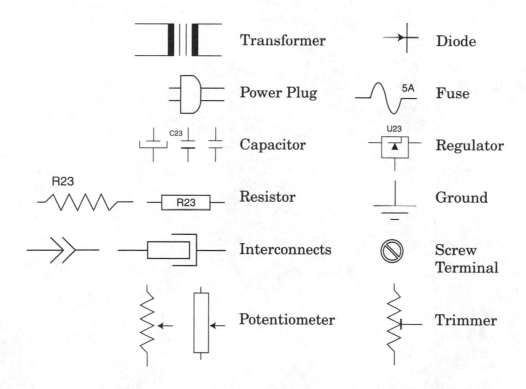

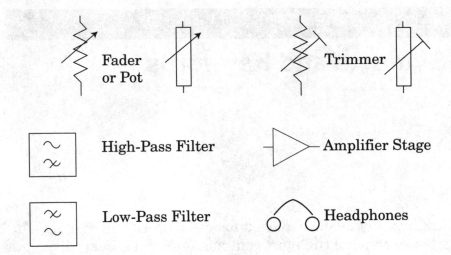

Fader or Pot

Trimmer

High-Pass Filter

Amplifier Stage

Low-Pass Filter

Headphones

Glossary

A The abbreviation for ampere.

Abrasion resistance The capability of a wire, cable, or material to resist surface wear.

AC *See* alternating current.

Accelerated aging A test that simulates long-time environmental conditions in a relatively short time.

ACR The difference between attenuation and cross talk, measured in dB, at a given frequency (also an acronym for *attenuation cross talk ratio*). ACR is an important characteristic in networking transmission to assure that a signal sent down a twisted pair is stronger at the receiving end of the cable than any of the interference signals imposed on that same pair by cross talk from other pairs.

AES/EBU Informal name of a digital audio standard established jointly by the *Audio Engineering Society* (AES) and the *European Broadcast Union* (EBU).

AF Audio frequency.

Air core transformers Transformers that are not gel filled.

Alloy A combination of two or more different polymers/metals usually combined to use different properties of each polymer metal.

Alternating current (AC) Electric current that alternates or reverses polarity continuously. The number of alternations per second are described as cycles (in hertz or Hz).

AM Amplitude modulation broadcasting.

Ambient   Normal conditions existing at a test or operating location prior to energizing equipment (such as ambient temperature).

American Wire Gauge (AWG)   A standard for expressing wire diameter. As the AWG number gets smaller, the wire diameter gets larger.

Ampacity   Current handling capability. The maximum current a conductor can carry without being heated beyond a safe limit.

Ampere   A standard unit of current. It is defined as the amount of current that flows when 1 volt of EMF is applied across 1 ohm of resistance. An ampere of current is produced by 1 coulomb of charge passing a point in 1 second.

Amplitude   The maximum value of a varying wave form.

Analog   The representation of data by continuously variable quantities.

Analog signal   An electrical signal that varies continuously, not having discrete values. Analog signals are copies or representations of other waves in nature. An analog audio signal, for instance, is a representation of the pressure waves that make up audible sound.

ANSI   American National Standards Institute.

ASTM   The American Society for Testing and Materials, a standards organization that suggests test methods, definitions, and practices.

Attenuation   The decrease in magnitude of a signal as it travels through any transmitting medium, such as a cable or some circuitry. Attenuation is measured as the logarithm of a ratio. It is expressed in decibels or dB.

Audio   A term used to describe sounds within the range of human hearing. Also used to describe devices designed to operate within this range (20 Hz to 20 kHz).

Audio frequency   Frequencies within the range of human hearing approximately 20 to 20,000 Hz.

AWG   *See* American wire gauge.

AWM   Appliance wiring material.

Backbone   The cable used to connect a multilevel distributed system with an intermediate system.

Backshell   A metal housing providing a shield through IDC connectors.

Balanced line   A cable having two identical conductors that carry voltages opposite in polarity and equal in magnitude with respect to ground. It is suitable for differential signal transmission.

Balun   A device for matching an unbalanced coaxial transmission line with a balanced two-wire system. Balun can also provide impedance transformation, such as 300-ohm balanced to 75-ohm unbalanced.

Bandwidth   The difference between the upper and lower limits of a given band of frequencies, expressed in hertz.

Baud rate   Unit of data transmission speed, meaning bits per second (500 baud = 500 bits per second).

Bel   A unit that represents the logarithm of the ratio of two levels.

Bend loss   A form of increased attenuation caused by (a) having an optical fiber curved around a restrictive radius of curvature or (b) microbends caused by minute distortions in the fiber imposed by externally induced perturbations.

Bend radius   The radius of curvature in which a flat, round, fiber optic, or metallic cable can bend without any adverse effects.

Binder   A tape or thread used for holding assembled cable components in place.

Bit   One binary digit.

Bit error rate   The number of errors occurring in a system per second, typically less than $10e^{-12}$.

Bits per second (bps)   The number of binary bits that can be transmitted per second. Examples are Mbps (mega is millions) and Gbps (giga is billions).

BNC   Bayonet Neil Concelman. A coaxial cable connector used extensively in video and *radio frequency* (RF) applications named for its inventor.

Booster   A device or amplifier inserted into a line or cable to increase the voltage. Transformers may be employed to boost AC voltages. The term booster is also applied to antenna preamplifiers.

Braid   A group of textile or metallic filaments interwoven to form a tubular flexible structure that can be applied over one or more wires. It can also be flattened to form a strap.

Braid angle   The angle between a strand of wire in a braid shield and the axis of the cable it is wound around.

Breakdown voltage   The voltage at which the insulation between two conductors fails and enables electricity to conduct or arc.

Breakout   The point at which a conductor or conductors are separated from a multiconductor cable to complete circuits at various points along the main cable.

Broadband   The technique used to multiplex multiple networks on a single cable without interfering with each other.

Buffer   A protective coating over an optical fiber.

Buffer amplifier   An amplifier that drives a low impedance load without adding voltage gain.

Bunch strand   Conductors twisted together with the same lay and direction without regard to geometric pattern.

Bus-bar wire   Uninsulated tinned copper wire used as a common lead.

Butyl rubber   A synthetic rubber with good insulating properties and that is highly flexible.

Byte   A group of adjacent binary digits (8 bits).

C   Abbreviation for capacitance and Celsius.

Cable   A group of individually insulated conductors twisted helically.

Cabling   The grouping or twisting together of two or more insulated conductors to form a cable.

Canadian Electrical Code (CEC)   Canadian version of the U.S. *National Electrical Code* (NEC).

Capacitance   The capability of a dielectric material between conductors to store energy when a difference of potential exists between the conductors. The unit of measurement is the farad. Cable capacitance is usually measured in *picofarads* (pF). Circuit capacitors are usually measured in *microfarads* (μF).

Capacitive cross talk   Cable cross talk or interference resulting from the coupling of the electrostatic field of one conductor upon another or more than one.

Capacitive reactance The opposition to alternating current due to the capacitance of a capacitor, cable, or circuit. It is measured in ohms and is equal to $\frac{1}{6.28}$ fC, where f is the frequency in hertz and C is the capacitance in farads.

Capacitor Two conducting surfaces separated by a dielectric material. The capacitance is determined by the area of the surfaces, the type of dielectric, and the spacing between the conducting surfaces.

Category The rating of a cable established by the *Telecommunications Industry Association* (TIA)/*Electronics Industry Association* (EIA) to indicate the level of electrical performance.

CATV Abbreviation for community antenna television.

CB Citizens band. A public-licensed radio band for communication.

CCTV Closed-circuit television.

Chrominance Signal The portion of a composite video signal that contains the color information.

Circuit A system of conducting media designed to pass an electric current.

Circular mil The area of a circle one one-thousandth of an inch (.001) in diameter. By knowing the circular mil area of various conductors, they can be used to determine the conductivity and gage size various combinations will produce.

Cladding A low refractive index material that surrounds the core of an optical fiber, causing the transmitted light to travel down the core. It protects against surface contaminant scattering using a layer of metal applied over another. Cladding is often chosen to improve conductivity or to resist corrosion.

Coaxial cable A cylindrical transmission line comprised of a conductor centered inside a metallic tube or shield separated by a dielectric material and usually covered by an insulating jacket.

Coil effect The inductive effect exhibited by a spiral-wrapped shield, especially above audio frequencies.

Color code A system of different colors or stripes used to identify values of components in a circuit or cable such as individual conductors or groups of conductors.

Component video   The unencoded output of a camera or video tape recorder whereby each red, green, and blue signal is transmitted down a separate cable. Component video systems most commonly use bundled coax as a transmission medium.

Composite video   The encoded output of a camera or video tape recorder whereby the red, green, blue, horizontal, and vertical sync are transmitted simultaneously down one cable.

Concentric stranding   A group of uninsulated wires twisted together containing a center core with subsequent layers spirally wrapped around the core with alternating lay directions to form a single conductor.

Conductivity   The capability of a material to enable electrons to flow, measured by the current per unit of voltage applied. It is the reciprocal of resistivity.

Conductor   A substance, usually metal, used to transfer electrical energy from point to point.

Conduit   A tube of metal or plastic through which wire or cable can be run. It is used to protect the wire or cable and, in the case of metal conduit, make it fireproof.

Connector   A device designed to enable an electrical flow from one wire or cable to a device on another cable. A connector will enable an interruption of the circuit or the transfer to another circuit without cutting any wire, cable, or other preparation.

Cord   A flexible insulated cable.

Core   The light-conducting central portion of an optical fiber with a refractive index higher than that of the cladding. It is the center of a cable construction and most often applies to a coaxial cable, where the core is the center conductor and the dielectric material applied to it.

Corona   The ionization of gasses around a conductor that results when the potential gradient reaches a certain value.

Coupling   The transfer of energy (without a direct electrical contact) between two or more cables or components of a circuit.

CPS   Cycles per second or hertz (Hz).

CPU   Central processing unit.

Cross talk   A type of interference caused by audio frequencies from one pair being coupled into adjacent pairs. The term is also used to describe coupling at higher frequencies.

CRT   Cathode ray tube.

CSA   Canadian Standards Association, the Canadian version of the Underwriters Laboratories.

Current carrying capacity   The maximum current a conductor can carry without being heated beyond a safe limit (ampacity).

Current loop   A two-wire transmit/receive interface.

D1   A component digital video recording format that conforms to the CCIR-601 standard and that uses 19-millimeter magnetic tape. (D1 is often used incorrectly to indicate component digital video.)

D2   A composite digital video recording format that uses 19-millimeter magnetic tape.

D3   A composite digital video recording format that uses ½-inch magnetic tape.

Daisy chain   A cable assembly with three or more termination areas.

dB   *See* Decibel.

DC   Direct current. The current is flowing in only one direction.

Decibel (dB)   A decibel is one-tenth of a bel and is equal to 10 times the logarithm of the power ratio, 20 times the log of the voltage ratio, or 20 times the log of the current ratio. Decibels are also used to express acoustic power, such as the apparent level of a sound. The decibel can express an actual level only when compared to some definite reference level assumed to be 0 dB.

Delay line   A transmission line or an equivalent device designed to delay a wave or signal for a specific length of time.

Derating factor   A multiplier used to reduce the current-carrying capacity of conductors in more adverse environments.

Dielectric   An insulating (nonconducting) medium when used in a signal-carrying design.

Dielectric breakdown   Any change in the properties of a dielectric that causes it to become conductive. Normally, it is the catastrophic failure of an insulation because of excessive voltage.

Dielectric constant   Also called permittivity. It is the property of a dielectric that determines the amount of electrostatic energy that can be stored by the material when a given voltage is applied to it. Actually, it is the ratio of the capacitance of a capacitor using the dielectric to the capacitance of an identical capacitor using a vacuum (which has a dielectric constant of 1) as a dielectric. It is a number that indicates the quality of a material to resist holding an electrical charge when placed between two conductors.

Dielectric heating   The heating of an insulating material when placed in a radio-frequency field caused by internal losses during the rapid polarization reversal of molecules in the material.

Dielectric loss   The power dissipated in a dielectric as the result of the friction produced by molecular motion when an alternating electric field is applied.

Dielectric strength   The voltage an insulation can withstand before it breaks down. Usually expressed as volts per mil.

Dielectric withstand voltage   The voltage that an insulating material can withstand before breakdown occurs.

Digital signal   An electrical signal that possesses two distinct states (on/off or positive/negative).

Direct current (DC)   An electrical current whose electrons flow in one direction only. It may be constant or pulsating as long as its movement is in one direction.

Dispersion   The cause of bandwidth limitations in an optical fiber. Dispersion causes a broadening of input pulses along the length of the fiber. Two major types are (a) mode dispersion caused by differential optical path lengths in a multimode fiber, and (b) material dispersion caused by a differential delay of various wavelengths of light in a waveguide material.

Distortion   Any undesired change in a wave form or signal.

Disturbed conductor   A conductor that receives energy generated by the field of another conductor or an external source. An example would be the quiet line.

Drain wire   A noninsulated wire in contact with parts of a cable, usually the shield, that is used in the termination to that shield and as a ground connection.

DVD   Short for Digital Versatile Disc, it is a high-density storage medium for data and video. Typically, it is 4.7 gigabytes.

E   Voltage (electromotive force). Similar to pressure in a water line.

Earth   British terminology for zero-reference ground.

EFB   Electronic field production. Video production for commercials, television shows, and other non-news purposes done outside the studio.

EIA   Electronic Industries Association (formerly RMA or RETMA).

Elastomer   Any material that returns to its original dimensions after being stretched or distorted.

Electromagnetic   Refers to the combined electric and magnetic fields caused by electron motion through conductors.

Electromagnetic coupling   The transfer of energy by means of a varying magnetic field.

Electron volt   A measure of the energy gained by an electron falling through an electric field produced by one volt.

Electrostatic   Pertaining to static electricity or electricity at rest, such as an electric charge.

Electrostatic coupling   The transfer of energy by means of a varying electrostatic field. Also known as capacitive coupling.

ELFEXT   Equal-level far-end cross talk (dB). A subtraction of attenuation from far-end cross talk (FEXT). By subtracting the attenuation, ELFEXT negates the effects of attenuation on the interference as it propagates down the cable, thus bringing it to an equal level.

EMF   Electromotive force (voltage).

EMI   Electromagnetic interference.

Energy   The capability of doing work, measured in many ways, such as in joules or watts.

Energy dissipation   A loss of energy from a system due to the conversion of work energy into an undesirable form, usually heat. The dissipation of electrical energy occurs when current flows through a resistance.

ENG   Electronic news gathering.

ETP   Abbreviation for a copper-refining process called electrolytic tough pitch. This process produces a conductor that is 99.95 percent pure copper, resulting in high conductivity.

EV   Electron volt.

f   Frequency.

Farad   A unit of capacity that stores one coulomb of electrical charge when one volt of electrical pressure is applied.

FAS   Fire Alarm and Signal cable, a *Canadian Standards Association* (CSA) cable designation.

FCFC   Flat conductor flat cable.

Feedback   Energy that is extracted from a high-level point in a circuit and applied to a lower level. Positive feedback reduces the stability of a device and is used to increase the sensitivity or produce oscillation in a system. Negative feedback, also called inverse feedback, increases the stability of a system as the feedback improves stability and fidelity.

Ferrous   Composed of or containing iron. A ferrous metal exhibits magnetic characteristics.

FEXT   Far-end cross talk (dB). It is cross talk induced on the pairs, measured at the "far" end of the cable.

Fiber   A single, separate optical transmission element characterized by core and cladding.

Fiber optics   Light transmission through optical fibers for communication and signaling.

Field   An area through which electric and/or magnetic lines of force pass.

Flat cable   Also referred to as planar or ribbon cable. It is any cable with two or more parallel conductors in the same plane encapsulated by insulating material.

Flat conductor   A conductor with a width-to-thickness ratio of 5 to 1 or greater.

Flat conductor cable   A flat cable with a plurality of flat conductors.

Flex life   The ability of a cable to bend many times before breaking.

Flexibility   The ability of a cable to bend in a short radius, lay flat, or conform to a surface, as microphone cables do.

Floating   Referring to a circuit that has no connection to ground.

FM   Frequency modulation where the frequency is shifted by the modulating signal, usually audio.

Frequency   The number of times a periodic action occurs in one second. Measured in hertz.

Frequency power   Normally, the 50 or 60 Hz of power available in residential areas.

Frequency response   The characteristic of a device denoting the range of frequencies over which it may be used effectively.

Gage   The physical diameter of a wire. As the *American Wire Gage* (AWG) number gets smaller, the wire diameter gets larger.

Gain   The increase of voltage, current, or power over a standard or previous reading, usually expressed in decibels.

Giga   One billion.

Gigahertz (GHz)   A unit of frequency equal to one billion hertz.

GND   *See* Ground.

Graded-index   A type of optical fiber in which the refractive index of the core is in the form of a parabolic curve, decreasing toward the cladding. This type of fiber provides high-bandwidth capabilities.

Ground   An electrical connection between a circuit and the earth. This also refers to a conductor connected to earth. In some instances, it can refer to a central metallic point designated as having "zero" potential.

Ground conductor   A conductor in a transmission cable or line that is grounded.

Ground loop   A completed circuit between shielded pairs of a multiple pair created by random contact between shields. An undesirable circuit condition in which interference is created by ground currents when grounds are connected at more than one point. This can also happen between chassis of different pieces of equipment. It is sometimes cured by single-end grounding techniques.

Ground potential   The potential of the earth. A circuit, terminal, or chassis is said to be at ground potential when it is used as a reference point for other potentials in the system.

H   A symbol designation for magnetic intensity. *Also see* Henry.

Harness   A flat cable or group of cables usually with many breakouts and with the wire ends prepared for termination or terminated to connectors and ready to install.

Henry   A practical unit of inductance that produces a voltage drop of one volt when the current changes at the rate of one ampere per second (abbreviated H).

Hertz (Hz)   The number of changes in polarity a signal makes in one second. An indication of frequency, it replaces the cycles per second measurement.

HF   *See* High frequency.

High frequency   The band from 3 to 30 MHz in the radio spectrum, as designated by the *Federal Communications Commission* (FCC).

Hum   A term used to describe the 60-cycle-per-second noise from some communications equipment. Usually, a hum is the result of undesired coupling to a 60-cycle source or a result of too many ground points causing ground loops. *See* grounding techniques.

Hz   *See* Hertz.

I   Symbol used to designate current.

I/O interconnection   Input/output interface to the outside world.

ICEA   Insulated Cable Engineers Association.

IDC   Insulation displacement contact.

IEEE   Institute of Electrical and Electronic Engineers.

IF   *See* Intermediate-frequency.

IFB   Interrupted feedback. A monitoring scheme often used in television where the program audio feed can be interrupted with directions, cues, or other information.

Impedance   The total opposition a circuit offers to the flow of alternating current or any other varying current at a particular frequency. It indicates the ideal transfer of signal from one piece of equipment to another. It is measured in ohms.

Impedance, high   Generally, the area of 25,000 ohms or higher.

Impedance, low   Generally, the area of 1 through 600 ohms.

Impedance match   A condition whereby the impedance of a particular circuit cable or component is the same as the impedance of the circuit, cable, or device to which it is connected.

Impedance matching transformer   A transformer designed to match the impedance of one circuit to that of another.

Inductance   The property of wire that stores electrical current in a magnetic field around the wire. By coiling wire, the effect can be intensified. It is measured in henrys.

Induction   The phenomenon of a voltage, magnetic field, or electrostatic charge being produced in an object by lines of force from the source of such fields.

Inductive cross talk   Cross talk resulting from the coupling of the electromagnetic field of one conductor upon another.

Inductive coupling   The process that occurs in a transformer to transfer energy from one winding to another.

Injection laser diode   Sometimes called the semiconductor diode. A laser in which the lasing occurs at the junction of n-type and p-type semiconductor materials.

Insertion loss   A measure of the attenuation of a cable or component by determining the output of a system before and after the device is inserted into the system.

Insulation   A material with good dielectric properties that is used to separate close electrical components, such as cable conductors and circuit components.

Interface   The region where two systems or a major and a minor system meet and interact with each other.

Interference   Disturbances of an electrical or electromagnetic nature that introduce undesirable responses into other electronic equipment.

Intermediate frequency   A frequency to which a signal is converted for ease of handling. It gets its name from the fact that it is an intermediate step between the initial and final conversion or detection stages.

Ionization   The formation of ions, which are produced when polar compounds are dissolved in a solvent and when a liquid, gas, or solid loses or gains electrons due to an electric current.

Ionization voltage   The potential at which a material ionizes and an atom gives up an electron.

IR   Infrared, which is a light wave that produces heat and light just below the visible light spectrum.

Isolation   The capability of a circuit or component to reject interference, usually expressed in dB.

Jacket   Pertaining to a wire or cable, the outer protective covering (it may also provide additional insulation).

Jumper   A short length of conductor or flat cable used to make a connection between terminals or around a break in a circuit or between circuit boards.

Kilo   One thousand.

kV   Kilovolt (1000 volts).

KVA   Kilovolt ampere.

kW   Kilowatt.

L   Abbreviation for inductance.

LAN   Local area network, which connects any number of users and is intended to serve a small area.

Laser   A coherent source of light with a narrow beam and a narrow spectral bandwidth (about 2 nanometers). It is an abbreviation for *light amplification by stimulated emission of radiation*. Lasers work as a result of resonant effects. The output of a laser is a coherent electromagnetic field. In a coherent beam of electromagnetic energy, all the waves have the same frequency and phase.

Lead dress   The placement or routing of wiring and component leads in an electrical circuit.

Lead-in   The cable that provides the path for *radio frequency* (RF) energy between the antenna and the receiver or transmitter.

Leakage   The undesirable passage of current over the surface of or through an insulator.

Level   A measure of the difference between a quantity or value and an established reference.

LF   Low frequency.

Light-emitting diode (LED)   A semiconductor device that emits incoherent light formed by the *positive/negative* (PN) junction. Light intensity is roughly proportional to electrical current flow (refers to the structure in the diode).

Line level   Refers to the output voltage level of a piece of electronic equipment. Usually expressed in decibels (0 dBv).

Line voltage   The potential value existing on a supply or power line.

Load   A device that consumes power from a source and uses that power to perform a function.

Long-wire antenna   Any conductor length in excess of half a wavelength. In a residential television installation, a horizontal run or unshielded lead-in acts as a long-wire antenna and introduces an additional signal on top of the regular antenna signal, causing ghosts.

Loss   When energy or a signal is lost without accomplishing useful work.

Lossy   Having poor efficiency.

Low frequency   A band of frequencies extending from 30 to 300 kHz in the radio spectrum, designated by the FCC.

Luminance signal   The portion of the composite video signal that represents the brightness or the black and white information.

M   Mutual inductance. The abbreviation for mega or 1 million. Indicates 1000 feet in the wire industry. An abbreviation for milli or one-thousandth.

mA   Milliampere (one-thousandth of an ampere).

Mbps   Megabits per second. The number of bits, in millions, transmitted per second.

Mega   Prefix meaning million.

Megahertz (MHz)   Unit of frequency equal to one million hertz (one million hertz per second).

μFd   Microfarad (one-millionth of a farad).

Mho   The unit of conductance equal to the reciprocal of the unit of resistance (ohm).

Micro   Prefix meaning one-millionth.

Microfarad   One-millionth of a farad (μf, μfd, mf, and mFd are common abbreviations).

Micro-microfarad   One-millionth of a microfarad (pf, μfd, mmf, mmfd are common abbreviations). Also, a picofarad (pf, pfd).

Micron   Millionth of a meter.

Microphonics   Noise caused by mechanical excitation of a system component. In a single-conductor microphone cable, for example, microphonics can be caused by the shield rubbing against the dielectric as the cable is flexed.

Mil   A unit of length equal to one-thousandth of an inch (.001 inch).

Milli   Prefix meaning one-thousandth.

Mode   A single electromagnetic wave traveling in an optical fiber.

Modem   A device that converts signals in one form to another form compatible with another kind of equipment.

Modulation   Altering the characteristics of a carrier wave to convey information. Modulation techniques include amplitude frequency and phase, plus many other forms of on-off digital coding.

Multiplex   A technique for putting two or more signals into a single channel.

Mutual capacitance   Capacitance between two conductors when all other conductors are connected together and grounded.

mV   Millivolt (one-thousandth of a volt).

mW   Milliwatt (one-thousandth of a watt).

Mylar   DuPont trademark for polyethylene terephtalate (polyester) film.

Nano   One-billionth.

Nanometer (nm)   One billionth of a meter.

Nanosecond   One billionth of a second.

Nibble   One half-byte (4 bits).

Noise   In a cable or circuit, any extraneous signal that tends to interfere with the signal normally present in or passing through the system.

NTSC (National Television Systems Committee)   An organization that formulated standards for the current American television system. It also describes the system of color telecasting that is used in Japan, Thailand, and parts of South America. NTSC television uses a 3.579545 MHz subcarrier whose phase varies with the instantaneous hue of the televised color and whose amplitude varies with the instantaneous saturation of the color. NTSC employs 525 lines per frame, 30 frames per second, and 59.94 fields per second.

OFHC   Oxygen-free, high-conductivity copper. It has 99.95 percent minimum copper content and an average annealed conductivity of 101 percent compared to standard copper.

Ohm   The unit of electrical resistance. The value of resistance through which a potential difference of 1 volt will maintain a current of 1 ampere.

Ohm's law   Stated $E = IR$, $I = E/R$, or $R = E/I$, the current (I) in a circuit is directly proportional to the voltage (E), and inversely proportional to the resistance (R).

Output   The useful power or signal delivered by a circuit or device.

Ozone   Extremely reactive form of oxygen, normally occurring around electrical discharges and present in the atmosphere in small but active quantities. In sufficient concentrations, it can break down certain rubber insulations under tension (such as a bent cable).

PAL (Phase Alternate Line)   PAL is a European color TV system featuring 625 lines per frame, with 25 frames and 50 fields per second. It is used mainly in Europe, China, Malaysia, Australia, New Zealand, the Middle East, and parts of Africa. PAL-M is a Brazilian color TV system with 525 lines per frame, with 30 frames and 60 fields per second.

Parallel circuit   A circuit in which the identical voltage is presented to all components, with the current being divided among the components according to the resistances or the impedance of the components.

Patchcord   A flexible piece of cable terminated at both ends with plugs. It is used for interconnecting circuits on a patchboard.

Peak   The maximum instantaneous value of a varying current or voltage.

Phase   An angular relationship between waves.

Phase shift   A change in the phase relationship between two alternating quantities.

Photo-detector   A receiver that converts light energy to electrical energy. The silicon photo diode is most commonly used for relatively fast speeds and has good sensitivity in the 0.75 to 0.95 micron wavelength region. *Avalanche photodiodes* (APD) combine the detection of optical signals with an internal amplification of the photo-current. Internal gain is realized through an avalanche multiplication of carriers in the junction region. The advantage in using an APD is its higher signal-to-noise ratio, especially at high bit rates.

Pickup   Any device capable of transforming a measurable quantity of intelligence (such as sound) into relative electrical signals (such as a microphone).

Pico   One-trillionth.

Picofarad   One billionth of a farad or a micromicrofarad. It is abbreviated as pF or mmF.

Pin-diode   A photodetector used to convert optical signals to electrical signals in a receiver.

Plug   A male housing with male or female contacts.

Point-to-point wiring   Wiring that consists of continuous conductors terminated at each end of the circuit.

Power   The amount of work per unit of time, usually expressed in watts and equal to the formula for power in watts. ($I^2 \times R$).

Power loss   The difference between the total power delivered to a circuit, cable, or device and the power delivered by that device to a load.

Power ratio   The ratio of power appearing at the load to the input power expressed in dB.

Precision video   Video coaxial cables having very tight electrical tolerances in impedance, the velocity of propagation, attenuation, and structural return loss. This video is used in high-quality applications such as live broadcasts in network studios and pre- or post-production facilities.

Propagation delay   Time required for a signal to pass from a device's input to its output.

Pulse   A current or voltage that changes abruptly from one value to another and back to the original value in a finite length of time. It is used to describe one particular variation in a series of wave motions.

R   Resistance.

RF   Radio frequency, usually considered to be frequencies ranging from 1 MHz to 3 GHz.

Rated temperature   The maximum temperature an electric component can operate at for extended periods without the loss of its basic properties.

Rated voltage   The maximum voltage an electric component can operate at for extended periods without undue degradation or safety hazards.

Reactance   A measure of the combined effects of capacitance and inductance on an alternating current. The amount of such opposition varies with the frequency of the current. The reactance of a capacitor decreases with an increase in frequency; the opposite occurs with an inductance.

Receptacle   A female housing with male or female contacts.

Reference edge   The edge of a cable or conductor from which measurements are made. It is sometimes indicated by a thread, an identification stripe, or a printing. Conductors are usually identified by their sequential position from the reference edge, with the number one conductor closest to this edge.

Reflection   The change in direction (or return) of waves striking a surface. For example, electromagnetic energy reflections can occur at an impedance mismatch in a transmission line, causing standing waves.

Refractive index   The ratio of light velocity in a vacuum to its velocity in the transmitting medium.

Registration   The alignment of one object in relation to another. In flat cables, it involves aligning conductors with contacts or solder pads. Registration is also simply called register.

Repeater   A receiver and transmitter combination used to regenerate an attenuated signal.

Resistance   In DC circuits, the opposition a material offers to current flow, measured in ohms. In AC circuits, resistance is the real component of impedance and may be higher than the value measured in DC.

Resonance   An AC circuit condition in which inductive and capacitive reactances interact to cause a minimum or maximum circuit impedance.

RFI   Ratio frequency interference.

RGB   Abbreviation for the three parts of a color video signal: red, green, and blue. It also refers to multicoaxial cables carrying these signals.

Ribbon cable   A flat cable made with parallel round conductors in the same plane. A ribbon cable is any cable with two or more parallel conductors in the same plane encapsulated by the insulating material. It is also referred to as planar or flat cable.

Ringing out   The process of locating or identifying specific conductor paths by means of passing a current through selected conductors.

RJ-45   Modular telecommunications connector.

RMS   Root mean square.

Routing   The path followed by a cable or conductor.

SAE   Society of Automotive Engineers.

Serial digital (SDI)   Digital information transmitted in serial form. It is often used informally to refer to serial digital television signals.

Series circuit   A circuit in which the components are arranged end to end to form a single path for the current.

Sheath   The outer protective covering of the wire and cable, which may also provide additional insulation.

Shield   A tape or braid (usually copper, aluminum, or other conductive material) placed around or between electric circuits, cables, or their components to prevent signal leakage or interference.

Shield effectiveness   The relative capability of a shield to screen out undesirable interference.

Shielded armored   Cables that require some sort of armor and shield. Types of shield include aluminum, aluminum/steel, gopher, and copper.

Signal   Any visible or audible indication that conveys information. A signal can also be the information conveyed through a communication system.

Single-ended   A circuit or transmission line that is grounded on one side.

Single mode fiber   A fiber waveguide in which only one mode is propagated. The fiber has a small core diameter of approximately 8 micrometers. It permits signal transmission at extremely high bandwidths and is generally used with laser diodes.

Sinusoidal   Varying in proportion to the sine of an angle or time function. An ordinary alternating current is sinusoidal.

Skew rays   A ray that does not intersect the fiber axis. Generally, it is a light ray that enters the fiber core at a very high angle.

Skin effect   The tendency of an alternating current to travel only on the surface of a conductor as its frequency increases.

Snake cable   Individually shielded and/or jacketed multipair audio cables. It is used in the connection of multichannel audio equipment.

SNR   Signal-to-noise ratio, commonly used interchangeably with *attenuation cross talk ratio* (ACR). It is the difference between attenuation and cross talk, measured in dB, at a given frequency. In networking transmission, it is important to assure that the signal sent down a twisted pair is stronger at the receiving end of the cable than any interference signals imposed on that same pair by cross talk from other pairs.

Sound Pressure Level (SPL)   The amount of sonic energy (sound pressure) in an area.

Source   The device (usually an LED or laser) used to convert an electrical information-carrying signal into a corresponding optical signal for transmission by an optical waveguide. It also refers to where a signal began or was generated.

Spectral bandwidth   The difference between wavelengths at which the radiant intensity of the illumination is half its peak intensity.

Spectrum   Frequencies that exist in a continuous range and have a common characteristic. A spectrum may be inclusive of many spectrums. For example, the electromagnetic radiation spectrum includes the light spectrum, radio spectrum, infrared spectrum, and so on.

Speed of light (*c*) 2.998×10^8 meters per second.

Standing wave   The stationary pattern of waves produced by two waves of the same frequency traveling in opposite directions on the same transmission line. The existence of voltage and current maxima and minima along a transmission line is a result of reflected energy from an impedance mismatch.

Standing wave ratio (SWR)   A ratio of the maximum amplitude to the minimum amplitude of a standing wave stated in current or voltage amplitudes.

Star quad   Term given to four-conductor microphone cables where the conductors are spiraled together. When connected in an x configuration, it greatly increases the common mode noise rejection.

Static charge   An electrical charge bound to an object or an unmoving electrical charge.

Surge   A temporary and relatively large increase in the voltage or current in an electric circuit or cable. Also called *transient*.

SVHS   Super VHS. A video format in which the two parts of the VHS video signal, the chrominance and luminance, are transmitted separately, providing a better picture resolution with less noise.

Sweep test   Testing the frequency response, or attenuation over the frequency, of a cable by generating a voltage whose frequency is varied through a given frequency range and observing or graphing the results.

Tensile strength   The pull stress required to break a bare wire.

Transducer   A device for converting mechanical energy to electrical energy or vice versa.

Transfer impedance   For a specified cable length, the transfer impedance relates to a current on one surface of a shield to the voltage drop generated by this current on the opposite surface of the shield. Transfer impedance is used to determine shield effectiveness against both the ingress and egress of interfering signals. Cable shields are normally designed to reduce the transfer of interference; hence, shields with a low transfer impedance are more effective than shields with a high transfer impedance.

Transmission line   An arrangement of two or more conductors, a coaxial cable, or a waveguide used to transfer signal energy from one location to another.

Triboelectric noise   Noise generated in a shielded cable due to variations in capacitance between the shield and conductors as the cable is flexed.

Twin lead   A transmission line having two parallel conductors separated by insulating material. Line impedance is determined by the diameter and spacing of the conductors, as well as the insulating material. It is usually 300 ohms for television-receiving antennas.

Twisted pair   Two lengths of insulated conductors twisted together.

UHF   Ultra high frequency, which is 300 to 3000 MHz.

UL   Underwriters Laboratories, a nonprofit organization that tests and verifies the construction and performance of electronic parts and equipment, including wire and cable.

Unbalanced line   A transmission line in which voltages on the two conductors are unequal with respect to ground. A coaxial cable is a common type of unbalanced line.

V   Volt.

VA   Volt-ampere. A designation of power in terms of voltage and current.

VHF   Very high frequency, which is 30 to 300 MHz, as designated by the FCC.

VHS   Video Home System, a trademark of Panasonic.

Video   Picture information in a television system.

VLF   Very low frequency, at 10 to 30 kHz.

Volt   A unit of electromotive force. Similar to pressure.

Voltage   The electrical potential of an electromotive force expressed in volts.

Voltage drop   The voltage developed across a component or conductor by the current flow through the resistance or impedance of the component or conductor.

Voltage rating   The highest voltage that may be continuously applied to a cable construction in conformance with standards or specifications.

Voltage standing wave ratio (VSWR)   The ratio of the transferring signal voltage as compared to the reflected signal voltage measured along the length of a transmission line.

VU   Volume units. A type of metering where the zero reference is to the nominal operating level of the device. Exemple: +4dB balanced out = OVU metered.

W   Watt or wattage.

Watt   A unit of electrical power.

Wave form   A graphical representation of a current's varying quantity. Usually, time is represented on the horizontal axis, and the current or voltage value is represented on the vertical axis.

Wavelength   The distance between a signal's positive peaks. As the frequency increases, and waves get closer together, the wavelength decreases.

X   Abbreviation for reactance.

XLR   Abbreviation for extra low resistance. A multipin audio connector (typically three pins) used in microphone, line-level, and snake cable connections.

Z   Abbreviation for impedance.

Test Questions

1. Describe or explain each of these terms:

   A. **Voltage** _____

   B. **Resistance** _____

   C. **Current** _____

   D. **Wattage** _____

   E. **Volt amps** _____

2. Show the symbol for each of these when calculating values:

   A. **Voltage** _____

   B. **Resistance** _____

   C. **Current** _____

   D. **Wattage** _____

   E. **Volt amps** _____

3. Describe how to find the approximate RMS output of a 117-volt amplifier without actually measuring it. (What would you look for in the calculations, keeping in mind that amplifiers are 80 percent efficient or better)?

4. Describe the difference between DC and AC currents.

5. What is the maximum power available continuously from a standard wall 117-volt outlet? Show your calculations. (Hint: what value are standard circuit breakers?)

6. What is the advantage of using 240 volts for large appliances like dryers?

7. Show the diagram and waveforms of a 240-volt feed. Label the lines and show the voltages of each line relative to neutral.

8. Why is AC power used instead of DC power for hydro-distribution?

9. Show the symbols for each item: transformer, power plug, diode, filter capacitor, regulator, and ground point.

10. Show a diagram for an XLR connector and label the pins. Also show the waveforms of each.

11. What is the advantage of a balanced signal over an unbalanced one?

12. Show a diagram of a tip ring sleeve connector and label the hot, cold, and ground.

13. What are the four basic types of meters?

14. What does each meter measure? (Describe the action of the meter display as well.)

15. If you have a 500-watt amplifier connected to a speaker, how much power is actually being converted to audio energy at best? (Keep in mind the efficiency of a speaker in general.)

16. What function do speakers and microphones have in common? (What are they and what do they do?)

17. Why do microphones require a ground at both ends of the connection?

18. What are the three main types of microphones?

19. What does a compressor do to the signal?

20. What does a noise gate do to the signal?

21. Show a diagram of an XLR connector and label the pins. Also show the waveforms on each.

22. Name the three main types of normalling connections.

23. Draw the three types.

24. What is different about the microphone connections in a patch bay?

25. Label the front layout of a patch bay starting from the top. Only a few channels in each strip are required. Start from the top row. Indicate what type of normalling is required on each row. Think of the signal flow through the mixer and recorder. Note that the first few are started for you.

26. What is the proper chemical for each type of cleaning job?

A. **Rubber pinch rollers** _____

B. **Ceramic capstans** _____

C. **Steel capstans** _____

D. **Steel rollers** _____

E. **Plastic rollers** _____

F. **Stationary heads** _____

G. **Rotary heads** _____

H. **Panels** _____

I. **Connectors** _____

J. **Pots and sliders** _____

K. **Amplifiers** _____

L. **Color wheels and mirrors** _____

27. Describe the procedure for cleaning a fixed-head tape recorder (tape path only).

28. Describe the procedure for cleaning a rotary-head tape recorder (digital and tape path only).

29. Describe the procedure for cleaning a lighting effects fixture.

30. List at least five of the major components in the tape path of a fixed-head tape recorder (show a diagram if desired).

31. List at least five of the major components in the tape path of a rotary-head tape recorder (show a diagram if desired).

32. List at least five of the components in an amplifier and which ones are dangerous to touch.

33. Why are the previous components dangerous to touch?

34. List at least five of the components in a light fixture and which ones are dangerous to touch?

35. Why arc these previous components dangerous to touch?

36. What is the power-up sequence of a sound system (what gets turned on last)?

37. Why is the power-up sequence done this way?

38. Describe the procedure for testing a speaker using a 9-volt battery. List the five basic steps that would confirm whether the fault is in the driver or cabinet, or toward the amplifier. Describe this process clearly for 10-point bonus.

A._____

B._____

C._____

D._____

E._____

Index

Note: Boldface numbers indicate illustrations or tables.

About the Author

Tom McCartney is owner and operator of T.M. Electronics, providing service to Toronto, Canada, area recording studios. Mr. McCartney both designs and builds audio and recording equipment, as well as working as a design consultant to other manufacturers. He also teaches a college-level studio operation and maintenance course, for which he developed this book.